TITANIC COLLECTIONS

VOLUME 1

FRAGMENTS OF HISTORY: THE SHIP

MIKE BEATTY, GEORGE BEHE, JOHN LAMOREAU, DON LYNCH, TREVOR POWELL AND KALMAN TANITO

TITANIC
COLLECTIONS

VOLUME 1

FRAGMENTS OF HISTORY: THE SHIP

Jacket Illustrations (Clockwise from top left): Luggage tag (George Behe collection); nurse's belt (Trevor Powell collection); 1911 brochure (Mike Beatty collection); third-class soup bowl (Mike Beatty collection); steward's badge (Trevor Powell collection); Turkish bath ticket (George Behe collection); Cadbury's tin (Kalman Tanito collection); White Star pin (Mike Beatty collection).

First published 2023

The History Press
97 St George's Place, Cheltenham,
Gloucestershire, GL50 3QB
www.thehistorypress.co.uk

© Mike Beatty, George Behe, John Lamoreau, Don Lynch, Trevor Powell and Kalman Tanito, 2023

The rights of Mike Beatty, George Behe, John Lamoreau, Don Lynch, Trevor Powell and Kalman Tanito to be identified as the authors of this work has been asserted in accordance with the Copyright, Designs and Patents Act 1988.

All rights reserved. No part of this book may be reprinted or reproduced or utilised in any form or by any electronic, mechanical or other means, now known or hereafter invented, including photocopying and recording, or in any information storage or retrieval system, without the permission in writing from the Publishers.

British Library Cataloguing in Publication Data.
A catalogue record for this book is available from the British Library.

ISBN 978 1 80399 333 1

Typesetting and origination by The History Press
Printed and bound in India by Thomson Press India Ltd

CONTENTS

Introduction	7
International Mercantile Marine Company	11
Harland & Wolff Ltd	13
White Star Line	16
Construction	24
Launch	30
Continued Construction	32
Leaving Belfast	45
Southampton	47
Publicity	50
Maiden Voyage	68
Disaster	99
Rescue	111
Aftermath	138
Author Biographies	172

INTRODUCTION

The basic facts of the story are well known. The White Star liner *Titanic* – the largest ship in the world – was scheduled to leave Southampton, England, on 10 April 1912 on the first leg of her maiden voyage. She was described as being 'practically unsinkable' and an oral tradition arose that ignored the qualifier 'practically' and promoted the belief among crewmen and the general public alike that the ship was *absolutely* unsinkable. In addition to her own passengers and crewmen, the *Titanic* was scheduled to carry a large number of travellers whose sailings on other liners had been cancelled due to a coal strike. Most passengers who were scheduled to sail on the brand-new ship eagerly let their friends and loved ones know about their travel plans, and many letters and postcards describing these plans were written to people living across the world.

At noon on 10 April, the *Titanic* gently eased away from the White Star dock in Southampton and began making her way down the channel towards the open sea. As the ship moved abreast of the moored liners *Oceanic* and *New York*, the sudden displacement of water created by the *Titanic*'s passing caused the *New York* to strain at her moorings. The lines connecting the *New York* to the *Oceanic* snapped like twine and the helpless vessel was relentlessly pulled out into the channel towards the huge passenger liner moving through the water nearby, her stern narrowly missing striking the *Titanic*'s port quarter. Several tugs wrestled with the smaller vessel as she slowly drifted forward along the *Titanic*'s port side, narrowly clearing the big liner's bows by a matter of feet. Despite this close call, the *Titanic* remained unscathed and, after a brief delay, she resumed her run for the open sea.

That evening, the *Titanic* paused at Cherbourg, France, to pick up additional passengers and mail, and at noon on 11 April she stopped again at Queenstown (now Cobh), Ireland, for that same purpose. After departing Queenstown, the *Titanic* turned her bows westward towards New York and headed out onto the vast expanse of the North Atlantic. Her passengers included some of the wealthiest and best-known personages in England and the United States, but – in addition to these luminaries – the great vessel carried a full complement of business and pleasure travellers, as well as a large number of European, Irish and Middle Eastern migrants, who were leaving their homelands in the hope of creating new lives for themselves in the New World. The *Titanic*'s passengers totalled 1,317, a number that was augmented by the vessel's 891 crewmen – most of whom hailed from the Southampton area and whose families depended upon the pay cheques the crewmen would receive from the White Star Line after they returned to England.

For the next three days, the *Titanic*'s maiden voyage proceeded normally and the ship's officers gradually increased her speed with each passing day, intending to better the maiden voyage crossing time achieved by her sister ship *Olympic* the previous year. The *Titanic*'s passengers enjoyed their brief respite from everyday life, taking advantage of this golden opportunity to socialise with friends and new acquaintances, listen to the orchestra, and partake in the excellent food and drink served in the ship's spacious dining rooms.

With the arrival of Sunday, 14 April, the *Titanic* began to receive wireless messages from other vessels warning of icebergs and field ice that stretched across the southern steamer track – the track the *Titanic* was traversing during her maiden trip to New York. The icefield ahead lay in a position that one officer calculated the *Titanic* would reach sometime around 11 p.m. that evening. Weather conditions and visibility remained ideal throughout the day, and at 7 p.m. three additional boilers were connected to the *Titanic*'s engines; this increased the vessel's speed to 22½ knots – the fastest speed she achieved during the entire voyage.

Darkness came and the *Titanic* steamed through the night with her speed undiminished, despite the presence of the icefield that was known to be situated directly in her path. Eleven o'clock came and went, and the *Titanic*'s lookouts were later heard to claim that they'd reported several 'early' iceberg sightings to the ship's officers beginning at about 11:15 p.m., but that the officers failed to slow the vessel down or take any other action to minimise the chances of a possible collision. Whether or not these claims were true, at 11:40 p.m. the officers' lack of prudent action finally caught up with them. The *Titanic* collided with an iceberg that opened six of her forward compartments to the sea.

After Captain Smith and ship designer Thomas Andrews determined that the *Titanic* was fatally damaged, the ship's officers began evacuating passengers from the vessel even though she was carrying an insufficient number of lifeboats to accommodate everyone on board. Two hours and forty minutes after the collision occurred, the ship upended, broke in half and sank beneath the waves, taking with her two-thirds of the passengers and crewmen who had entrusted their lives to the White Star Line. The 712 people who were lucky enough to find places in lifeboats or be plucked from the water were picked up by the Cunard liner *Carpathia*, which landed them in New York three days later.

During the days following the disaster, several ships were sent out from Halifax, Nova Scotia, to retrieve the floating bodies of as many *Titanic* victims as they could find. Meanwhile, the United States Senate began its inquiry into the reasons why the disaster occurred in the first place. In subsequent weeks, the British government conducted its own inquiry into the disaster, and during the ensuing months and years many memorials to individual *Titanic* passengers and crewmen, as well as memorials to groups of victims such as the ship's engineers and bandsmen, were erected in various places throughout the world.

As time passed, interest in the *Titanic* disaster gradually became dormant, but newspapers throughout the world nevertheless revived the ill-fated vessel's memory during the month of April by interviewing any survivors who happened to be living within the newspapers' home districts. The 1955 publication of Walter Lord's book *A Night to Remember*, the 1985 discovery of the *Titanic* wreck site by Dr Robert Ballard and James Cameron's 1997 blockbuster film all played their own part in keeping the ship's memory alive. Museum exhibits of *Titanic* artefacts continue to drive home the stark fact that the *Titanic* was a real ship and that 1,496 men, women and children travelling on board the great liner never lived to reach their final destinations.

Not all *Titanic* artefacts reside in museums, however, because many individual researchers collect such artefacts and documents to supplement their own interest in the great passenger liner. A few such artefacts are occasionally lent to museums for public display, but most items remain 'hidden' in private collections and often disappear from the historical record, simply because the secretive nature of many collectors causes them to relish the fact that they possess items nobody else will ever see.

This latter category of collectors does not include this book's co-authors, all of whom have collected *Titanic*-related material for most of their lives and take pleasure in sharing their acquisitions with researchers all over the world. Indeed, this is the sole reason for the existence of the book you now hold in your hands – so that we can share rare *Titanic* items with like-minded researchers who may never have been aware that these items existed.

We hope the rare artefacts and documents illustrated within the following pages will augment future historians' study of the *Titanic* disaster and will help people remember that real human beings lost their lives when this real ship went down in the North Atlantic on 15 April 1912.

George Behe
Michael Beatty
John Lamoreau
Don Lynch
Trevor Powell
Kalman Tanito

INTERNATIONAL MERCANTILE MARINE COMPANY

▲ **Stock Certificate**
One of the earliest International Mercantile Marine Company stock certificates, number 6, dated 1 December 1902, for Thomas W. Joyce, for 49,890 shares at a value of $100 each. The IMM was a holding company that controlled subsidiary corporations, one of which was the White Star Line. (Kalman Tanito collection.)

▲ **Stock Certificate**
A later stock certificate for the International Mercantile Marine Company. (George Behe collection.)

FRAGMENTS OF HISTORY: THE SHIP 13

HARLAND & WOLFF LTD

▶ **Register of Shareholders**
Harland & Wolff's Register of Shareholders from the 1940s onwards. Harland & Wolff was the Belfast shipbuilding firm that constructed the White Star Line's *Olympic* and *Titanic*. (Kalman Tanito collection.)

▲ **Builder's Plaque**
A Harland & Wolff builder's plaque, similar to the one on the *Titanic*. (Kalman Tanito collection.)

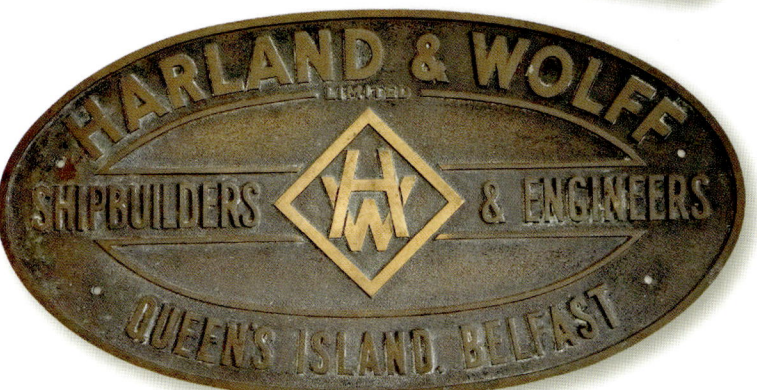

▲ **Builder's Plaque**
A later Harland & Wolff builder's plaque. (Kalman Tanito collection.)

◀▲ **Lantern Slides**
Magic lantern slide showing the new Arrol Gantry built at the Harland & Wolff shipyard in 1909. Three slipways were removed and two new ones with the gantry were built in their place to accommodate the size of the *Olympic* and *Titanic*. (Mike Beatty collection.)

WHITE STAR LINE

▲ **Postcards**
Views of the White Star offices in Dublin and London where many passengers purchased tickets to travel to America on board the company's transatlantic steamers. (George Behe collection.)

▶ **White Star Line**
The lease agreement for the White Star Line's Liverpool offices at 30 James Street, complete with floor plans, signed by Harold Sanderson. (Kalman Tanito collection.)

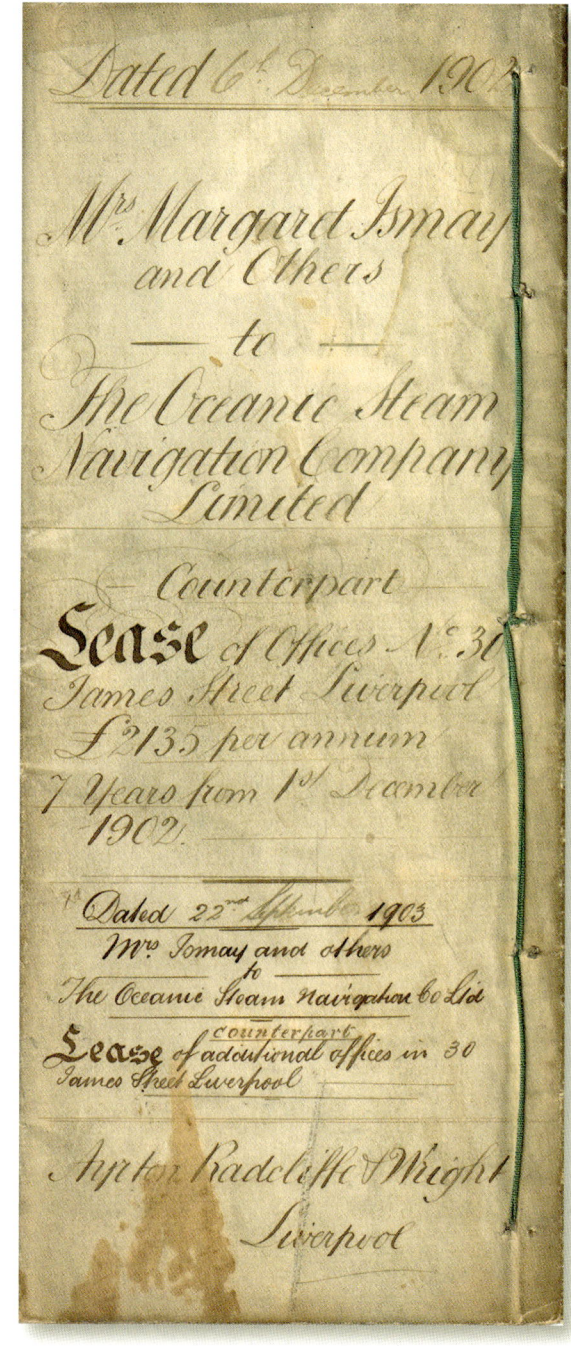

FRAGMENTS OF HISTORY: THE SHIP   17

▲ **Advertisement**
A large period cardboard cut-out White Star Line advertisement. (Trevor Powell collection.)

▲ **White Star Line Pin**
A small celluloid advertising pin from the period. (Mike Beatty collection.)

▲ **Button**
This button bearing the White Star burgee is of the same pattern used on uniforms of the White Star Line's navigating officers in 1912. (George Behe collection.)

▲ **Steward's Badge**
A c. 1911 brass and enamel White Star Line first-class steward's badge. 'Saloon' indicates that it was worn by a dining saloon steward, each of whom was issued a number that was used for identification purposes on board. During the body recovery process, many stewards were identified based on the badge number found on their person. (Trevor Powell collection.)

▲ **Steward's Cap**
A *c.* 1911 White Star Line wool steward's cap with a brass and enamel star, outfitted by J.J. Rayner & Sons of Liverpool. Although identical examples were worn by stewards on the *Titanic*, this example was worn by a crewman serving aboard the *Olympic* in 1912. (Trevor Powell collection.)

FRAGMENTS OF HISTORY: THE SHIP 19

▲ **Nurse's Belt**
A leather and enamel White Star Line nurse's belt, *c.* 1910. (Trevor Powell collection.)

◀ *Adriatic* Passenger List

A first-class passenger list of SS *Adriatic*, dated 30 December 1908. The list mentions both the *Titanic* and the *Olympic*, describing them as 'THE LARGEST STEAMERS IN THE WORLD – BUILDING'. The *Adriatic*'s captain, Edward J. Smith, would later captain the *Titanic* on her maiden voyage and *Adriatic*'s surgeon, W.F.N. O'Loughlin, was destined to serve on the *Titanic* as well. Also present on the *Adriatic* during this voyage were future *Titanic* passengers George and Eleanor Widener. (John Lamoreau collection.)

FRAGMENTS OF HISTORY: THE SHIP

◀ **Carl Eriksson Brochure**
The White Star Line advertised heavily to Swedish migrants. This 1910 brochure shows all the current ships that migrants could travel on and features the *Titanic* and *Olympic* under construction. (Mike Beatty collection.)

▼ **Advertising Postcard**
It was probably in 1910 that this postcard was printed in Sweden advertising the White Star Line's stable of thoroughbred passenger liners; the card mentions that the Line's two newest vessels, the *Olympic* and *Titanic*, were now under construction. (George Behe collection.)

▼ **The Shipbuilder**
A very rare copy of the special *Titanic* and *Olympic* edition of *The Shipbuilder* from 1911.
Later that same year they published a souvenir edition with a green hardcover. (Mike Beatty collection.)

FRAGMENTS OF HISTORY: THE SHIP   23

▲ Painting
Oilette of the *Olympic* by Fred Hoertz (1889–1978). (Kalman Tanito collection.)

CONSTRUCTION

◀ **Postcard**
The *Titanic* (left) and her elder sister *Olympic* were both constructed at the Harland & Wolff shipyard in Belfast, Ireland. (George Behe collection.)

◀ **Postcard**
This postcard shows a stern view of the *Titanic* (right) and *Olympic* as their construction continues at Harland & Wolff. (George Behe collection.)

FRAGMENTS OF HISTORY: THE SHIP 25

▶ **Postcard**
The two sister ships on 20 October 1910, just prior to the *Olympic*'s launching. (George Behe collection.)

▶ **Postcard**
A postcard view of some of the Harland & Wolff workmen who constructed the *Olympic* and *Titanic*. (George Behe collection.)

HARLAND & WOLFF'S NORTH YARD; WITH THE "TITANIC" ON THE WAYS

▶ **Sailing Schedule**
A sailing schedule for the White Star Line's upcoming 1911 season. The *Olympic* and *Titanic* are both listed as still being under construction. (George Behe collection.)

◀◀ **Top Image: Postcard**
This view of the Harland & Wolff shipyard shows the *Titanic* still under the huge gantry as work continues apace on the great liner's construction. Her launch is now just a few days away. (George Behe collection.)

Bottom Images: *Olympic* Launch
A spectator out on the water captured several amateur photos of the *Olympic* launch on 20 October 1910. The white stern of *Olympic* can be seen in the left side of the gantry while the black hulk of *Titanic* is on the right side of the photograph. Several minutes later, the photographer captured the *Olympic* in the water. The *Titanic* would have to wait until May of the next year for her own launching. (Mike Beatty collection.)

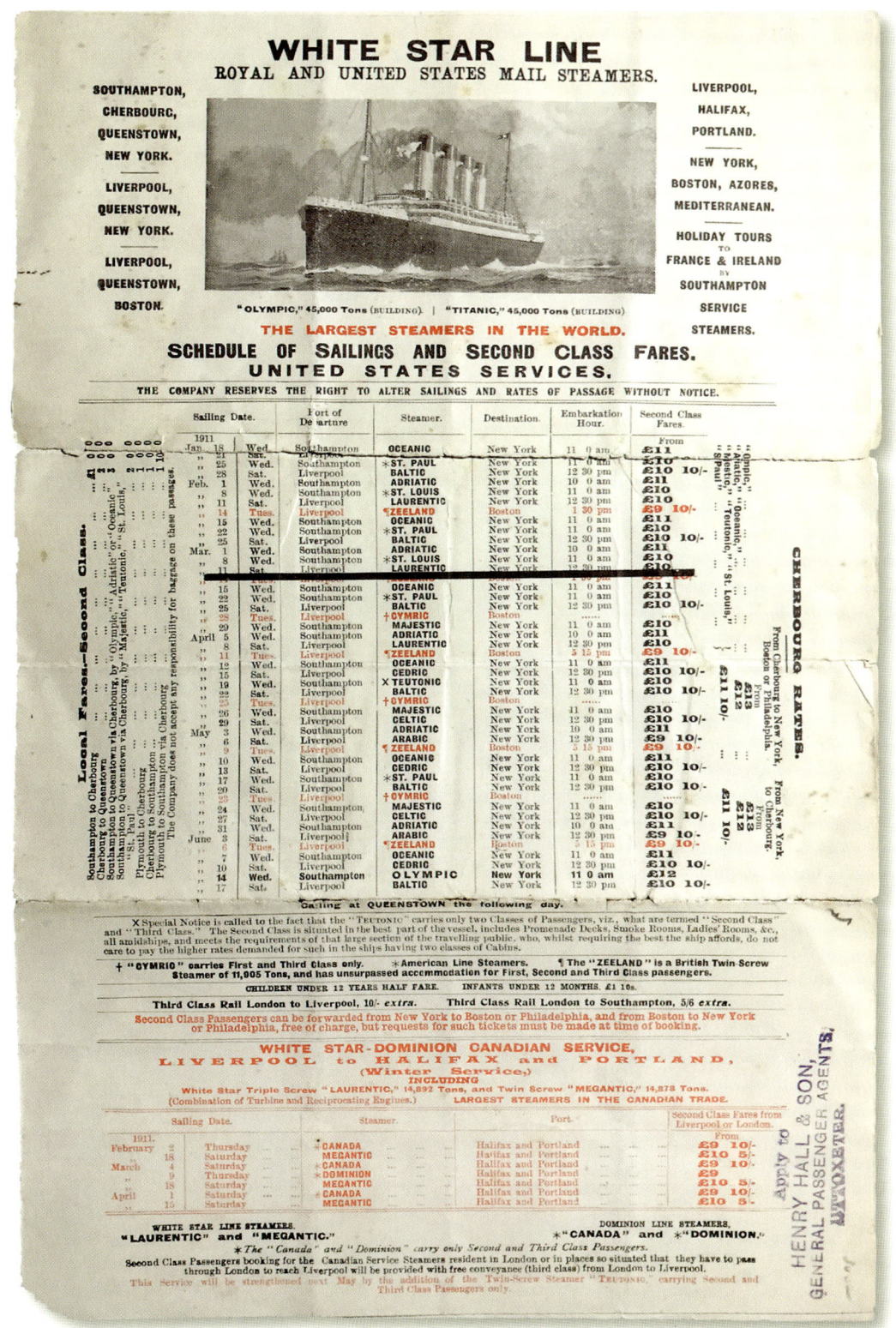

THE LARGEST STEAMERS IN THE WORLD
FACTS ABOUT THE
WHITE STAR LINE'S
NEW TRIPLE-SCREW STEAMERS
"OLYMPIC" "TITANIC"

Tonnage, registered	45,000
Tonnage, displacement	66,000
Length over all	882 feet, 6 inches
Breadth over all	92 feet, 6 inches
Breadth over boat deck	94 feet, 0 inches
Height from bottom of keel to top of Captain's house	105 feet, 7 inches
Height of funnels above casing	72 feet, 0 inches
Height of funnels above boat deck	81 feet, 6 inches
Distance from top of funnel to keel	175 feet, 0 inches
Number of steel decks	11
Number of watertight bulkheads	15
Engines	Combination Turbine and Reciprocating
Anchors, each	15½ tons
Anchor Cable Links, each	175 pounds
Rudder	100 tons
Rivets used, three millions, weighing	1,200 tons
Wing Propellors, each	38 tons
Center Propellor	22 tons
Sidelights in each ship	2,200
Crew carried	890
Passenger capacity	2,500

Sports Decks and Spacious Promenades
Commodious Staterooms and Apartments en suite
Cabins de Luxe with Bath
Squash Racquet Courts
Turkish and Electric Bath Establishments
Salt Water Swimming Pools
Glass-Enclosed Sun Parlors
Verandah Cafés
French a la carte Restaurants
Grand Dining Saloons
Electric Elevators in First and Second Class

WHITE STAR LINE OFFICES
Nine Broadway, New York

New York . Pier 62, North River
Boston . 84 State St., India Bldg.
Chicago Cor. Washington and La Salle Sts.
Halifax . . . 159 Hollis Street
Minneapolis 121 South Third Street
Montreal 118 Notre Dame Street, W.
New Orleans 219 St. Charles Street
Philadelphia . 1319 Walnut Street

Portland, Me. . . 1 India Street
Quebec . . 53 Dalhousie Street
San Francisco . 319 Geary Street
St. Louis . . . 900 Locust Street
Seattle . . 619 Second Avenue
Toronto . . 41 King Street, East
Washington, D. C. 1306 F St., N.W.
Winnipeg, Man. 333 Main Street

—OR TO—

C1819 50M-10-29-11

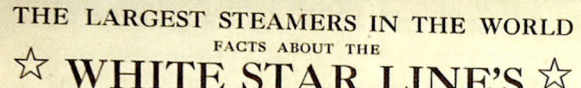

WHITE STAR LINE
THE LARGEST AND FINEST STEAMERS IN THE WORLD
OLYMPIC
45,000 TONS
TITANIC
45,000 TONS

Some Interesting Views

▲ **Postcard**
On 29 April 1911, a photographer at Noah Hingley & Sons Ltd in Dudley, England, snapped a photograph of *Titanic*'s newly painted centre anchor. The next day a long team of horses hauled the wagon to the railway station so that the anchor could be delivered by train to Harland & Wolff. (George Behe collection.)

◀◀ **1911 Foldout Brochure**
This large brochure folds out to over 3ft and reveals a fantastic cutaway view of the two sister ships.
The reverse side illustrates some of the two vessels' first-class public rooms and cabin options. (Mike Beatty collection.)

LAUNCH

▲ Postcard
At 12:15 p.m. on 31 May 1911 – a bright and sunny day for launching – the *Titanic* begins to slide out from underneath Harland & Wolff's great gantry into the waiting waters of the River Lagan. (George Behe collection.)

LAUNCH OF WHITE STAR R. M. S. "TITANIC" 31ST MAY 1911. AT BELFAST.

▶ Postcard
The *Titanic*'s stern gradually picks up speed and dips into the water for the very first time as the great vessel slides aft down the ways. (George Behe collection.)

▶ Postcard
At 12:16 p.m., the *Titanic* found herself afloat at last as drag chains attached to her hull began to arrest her backwards movement. The launch process had taken just sixty-two seconds, and the great vessel would now be towed to the fitting-out wharf so that the engines and heavy machinery could be lowered in as the construction process continued. (George Behe collection.)

CONTINUED CONSTRUCTION

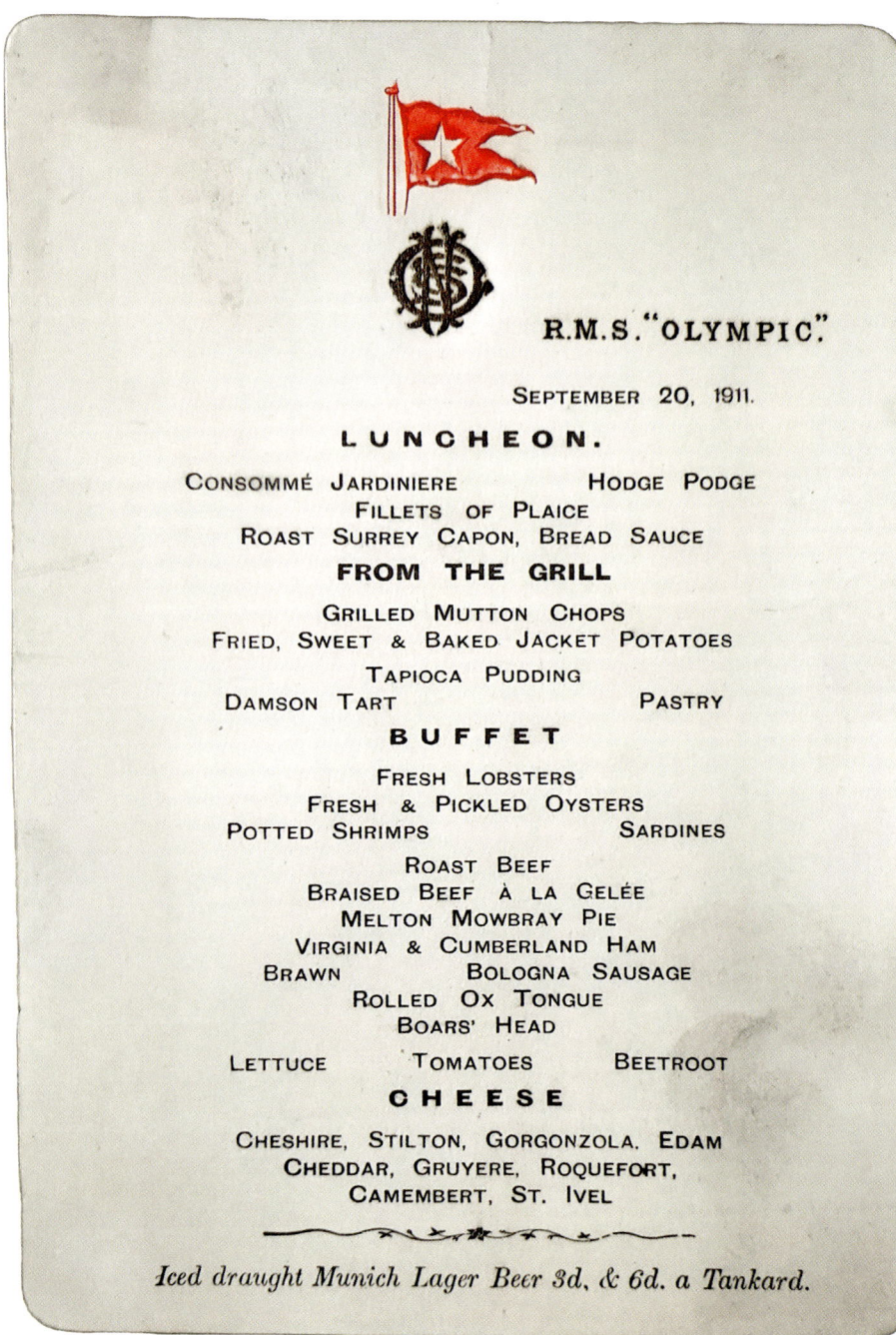

▲ **Maiden Voyage Advertisement**
On 20 September 1911, the *Olympic* collided with HMS *Hawke*, which caused work on the *Titanic* to be delayed while many of her workers were detached to deal with *Olympic*'s repairs. This disruption in routine caused *Titanic*'s maiden voyage to be postponed from 20 March to 10 April 1912, and an advertisement for that revised sailing date appeared in the 22 February 1912 issue of *The London Times*. (Mike Beatty collection.)

◀ **Menu**
A first-class luncheon menu from the *Olympic*, dated from the day of the HMS *Hawke* collision. Several crewmembers who would later serve aboard the *Titanic* were on board that day. (Trevor Powell collection.)

FRAGMENTS OF HISTORY: THE SHIP

▶ **Postal Cover**
When repairs to the *Olympic* caused the *Titanic*'s March maiden voyage to be postponed until 10 April, mail originally scheduled to be transported by the *Titanic* in March was transferred to other ships to facilitate timely delivery. The postal cover pictured here was one of the items that narrowly escaped being transported in the *Titanic*'s mail room – the only reason it still exists. (George Behe collection.)

▶ **Postcard**
After the *Olympic*'s collision repairs were completed, the White Star Line advertised her return to the transatlantic service and added that the brand-new *Titanic* would soon be entering that service as well. (George Behe collection.)

◀ **Maiden Voyage Postcard**
This rare postcard advertises the date of the maiden voyage of the *Titanic*. It was likely published just a few weeks beforehand. (Trevor Powell collection.)

◀ **Postcard**
The photograph on this postcard is usually misidentified as depicting the *Titanic*'s entire complement of officers. In truth, the photograph was snapped on board the *Olympic*, and only three of these officers – Captain Edward J. Smith (front, second from right), Chief Purser Hugh McElroy (left rear) and First Officer William Murdoch (front right) – were later transferred to the *Titanic* for her maiden voyage. (George Behe collection.)

▲ **Newspaper Advertisements**

In March 1912, the White Star Line began advertising the fact that the *Olympic* and *Titanic* would be working in tandem on the Southampton–New York route as soon as the newer vessel's construction was completed. The planned date for the second half of *Titanic*'s 10 April maiden voyage – 20 April – was also promoted with great optimism. (George Behe collection.)

▲ *Titanic* and *Olympic* Lithograph
This large advertising lithograph by famed maritime artist Montague Black features the *Olympic* in the foreground passing the *Titanic* in the distance and was the only advertising to show them individually and not as one image representing both ships. These were printed by the Liverpool Printing Company and would have hung in White Star Line offices. These illustrations originally had a green border with wording advertising both ships, but after the sinking most had the green border removed and were reframed so they could still be used. Only a handful survive today. (Mike Beatty collection.)

FRAGMENTS OF HISTORY: **THE SHIP** 37

▶ Postcard

The *Titanic* sits in the dry dock as work on her construction continues. By the time March 1912 arrived, the engines and boilers had been successfully lowered into the vessel, after which the ship's four huge funnels were installed. (George Behe collection.)

▶ Postcard

On 2 March 1912, the *Olympic* (left) was forced to return to the Harland & Wolff shipyard yet again so that repairs could be made to a damaged propeller. Meanwhile, workmen raced to complete the *Titanic*'s construction before the advertised date of her maiden voyage arrived. (George Behe collection.)

▲ **Decking**
This piece of wooden decking was purchased in 1935 when the *Olympic* was scrapped. Identical decking was used on the *Titanic*. (John Lamoreau collection.)

▲ ***Titanic* Carpet Fragment**
While *Titanic*'s construction continued in Belfast, a scrap of excess carpet was removed from the ship by steward Fred Ray, who wanted to show his wife the richness of the *Titanic*'s appointments. Ray (who would survive the sinking) afterwards used the carpet fragment to stuff a footstool he was constructing for his family's use. Decades later, Ray tossed the worn-out footstool out into his back garden prior to disposing of it altogether, but when the stool broke open, he recognised the carpet fragment that fell out of it. Realising its historical significance, Ray donated the carpet to the Titanic Historical Society, which later offered several 1in squares of the carpet in exchange for new additions to the society's own collection. (George Behe collection.)

FRAGMENTS OF HISTORY: THE SHIP   39

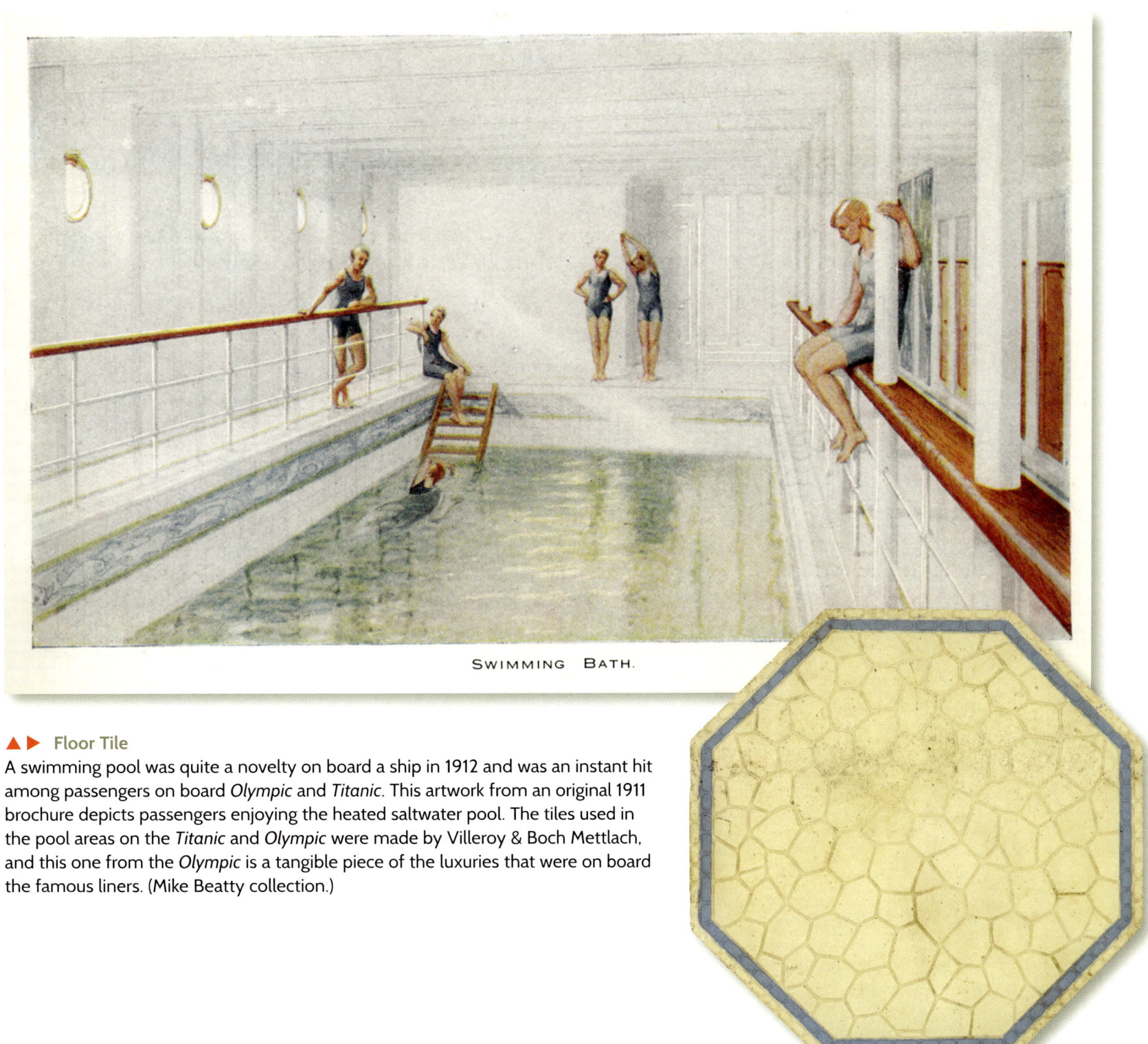

SWIMMING BATH.

▲▶ Floor Tile

A swimming pool was quite a novelty on board a ship in 1912 and was an instant hit among passengers on board *Olympic* and *Titanic*. This artwork from an original 1911 brochure depicts passengers enjoying the heated saltwater pool. The tiles used in the pool areas on the *Titanic* and *Olympic* were made by Villeroy & Boch Mettlach, and this one from the *Olympic* is a tangible piece of the luxuries that were on board the famous liners. (Mike Beatty collection.)

◀ **Floor Tiles**

Although these linoleum tiles came from the *Olympic*, identical tiles were used on the *Titanic*. These alternating green and white linoleum tiles appeared in many public and crew spaces on board the two vessels, and when the *Olympic* was scrapped in 1935, these tiles were purchased by the Smith and Walton paint factory in Haltwhistle, Northumberland, England. (John Lamoreau collection.)

▶▶ **Grand Staircase Oak Moulding**

When the *Olympic* was scrapped, these two wood fragments were taken from her grand staircase, which was identical to the *Titanic*'s. *(Top)*: This almost 2ft length of oak carved moulding was covered with multi-layers of paint which were later stripped off to reveal its original oak finish. *(Bottom)*: This 9in length of carved oak moulding is still covered with multi-layers of paint that were applied over the decades. (John Lamoreau collection.)

FRAGMENTS OF HISTORY: THE SHIP   41

◀ **Carved Moulding**
This sycamore carved moulding came from the first-class section of the *Olympic* and was salvaged when that vessel was scrapped in 1935. (John Lamoreau collection.)

▶ **First-Class Brochure**
This rare oversized 1911 brochure features large photos and descriptions of the amenities offered in first class on the new *Olympic* and *Titanic*. (Mike Beatty collection.)

FRAGMENTS OF HISTORY: **THE SHIP** 43

▲ **Titanic Reverse Glass Painting**
This pair of 13in square reverse glass paintings feature the *Titanic* and *Olympic* and are titled 'Largest Steamer in the World'. The pair of wall hangings was sold as a publicity souvenir before the sinking occurred. (Mike Beatty collection.)

▲ **Oversized Postcard**
By 1 April 1912, the *Titanic*'s screen of windows on A Deck had been installed, permitting her to be easily distinguished from her elder sister *Olympic*. The great vessel was now waiting to begin her trials, but high winds forced a postponement of the trials until 2 April. (George Behe collection.)

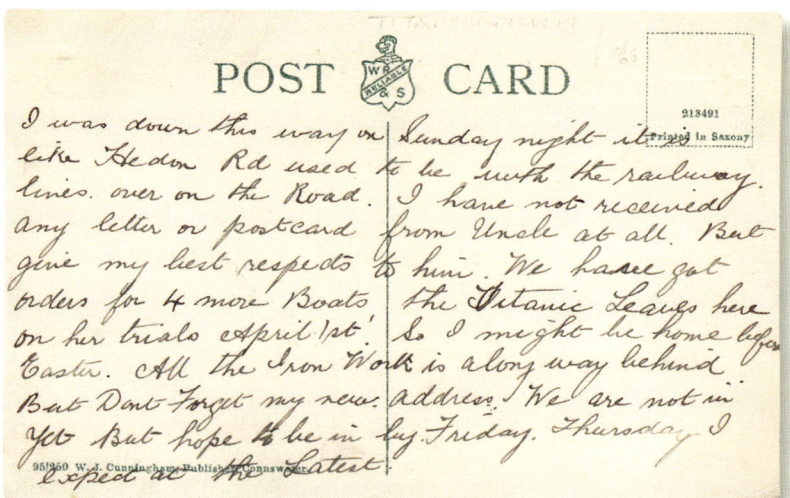

▲ **Shipyard Worker Postcard**
A Harland & Wolff shipyard worker writes that the *Titanic* will soon be leaving for her sea trials. (Mike Beatty collection.)

LEAVING BELFAST

▶ **Postcards**

Two postcard views of the *Titanic* beginning her sea trials on 2 April 1912. After the completion of these trials the ship proceeded to Southampton and arrived in that port city during the early hours of 4 April. (George Behe collection.)

The largest vessel in the world, the new White Star Liner "Titanic" leaving Belfast, April 3rd, 1912.   Length, 882 ft. 9 in. ; breadth, 92 ft. 6 in. 45,000 tons gross register ; 66,000 tons displacement.   Accommodation, 2,500 passengers and a crew of 860.
WALTON, PUBLISHER, BELFAST.

FAREWELL TO BELFAST.
DEPARTURE OF THE LARGEST VESSEL IN THE WORLD,
THE WHITE STAR LINER "TITANIC," 46,328 TONS. APRIL 3rd, 1912.

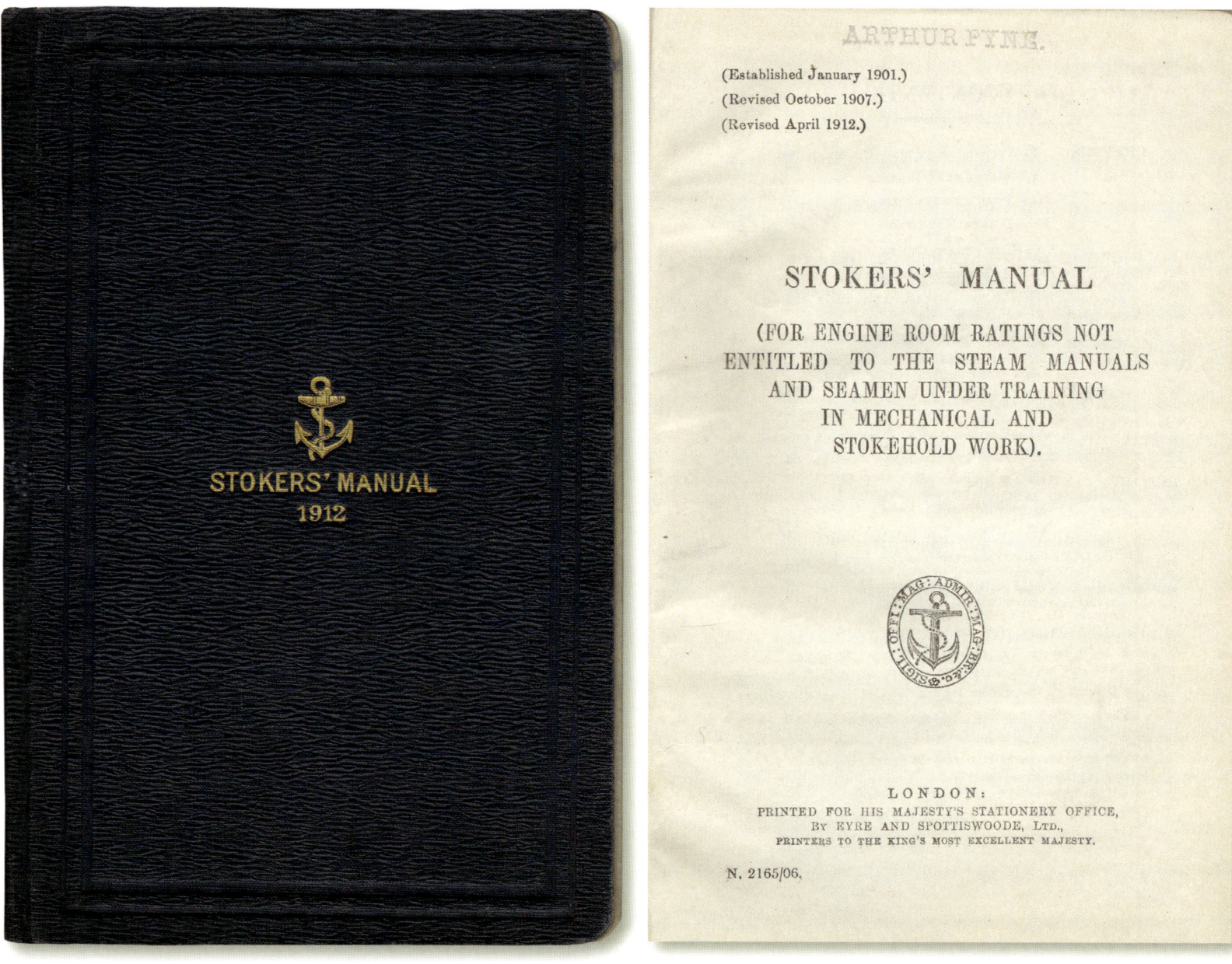

▲ Stokers' Manual
Copies of this rare April 1912 edition of the stokers' training and reference book were likely carried by several of the newer stokers working on board who were still learning their jobs. Ironically, many of the stokers could not read, or even sign their own names. (Mike Beatty collection.)

SOUTHAMPTON

▲ Postcard
The *Titanic* sits at the Southampton dock on the morning of 4 April 1912. Harland & Wolff workers continue to work on the great vessel's interiors, while provisions and coal are taken aboard in preparation for the 10 April maiden voyage. (George Behe collection.)

◀ **Postcard**
The *Titanic* rests peacefully at the dock in Southampton. During the ensuing week hectic activity on board the great vessel was non-stop as Harland & Wolff workers feverishly tried to complete their work of installing fittings, painting rooms, and finishing the thousand-and-one other tasks that needed to be completed before the *Titanic* could leave the dock on 10 April. (George Behe collection.)

▲ **Photograph**
Titanic was photographed on 7 April 1912 as she sat at dockside in Southampton, her maiden voyage just three days away. 'She was so much larger than one even expected,' one visitor to the ship later wrote. 'She looked so solidly constructed, as one knew she must be, and her interior arrangements and appointments were so palatial that one forgot now and then that she was a ship at all. She seemed to be a spacious regal home of princes.' (George Behe collection.)

PUBLICITY

Postcards

A selection of pre-maiden voyage postcards advertising the brand-new *Titanic*. Cards like these were sold on board the vessel for use by her passengers as well as sold on shore and in hotels to promote the White Star Line and its ships. The central card opposite, bearing the vessel's woven-in-silk likeness, was undoubtedly more expensive than the others, which accounts for its rarity today. (George Behe collection.)

▲ Postcard
This pre-maiden voyage card shows an artist's surprisingly primitive depiction of what the *Titanic* would look like after she was finally put into service. Although the artwork on many post-disaster memorial postcards was hastily executed and sub-par, this is the only pre-maiden voyage card we're aware of that utilises such amateurish artwork. (George Behe collection.)

▶ **Postcards**

Two views of the *Olympic*'s first-class smoking room. On the *Titanic* this room was a male stronghold where good cigars, good beverages, good conversation and good card games held sway. 'I sat in the big carved mahogany settee, with deep, wide springy leather upholstering, and toasted my feet at the big coal fire that blazed in a fireplace worthy of a king's palace,' a Southampton visitor to the *Titanic* wrote later. 'Over the fireplace was a beautiful sea picture by Mr Norman Wilkinson. The settee formed two horns on either side of the fire, and a dozen folk could sit in this settee in comfort.' (George Behe collection.)

1193 C. R. Hoffmann, Southampton. R.M.S. OLYMPIC Length 900 Ft. Breadth 99 Ft. 46,359 Tons
FIRST CLASS VERANDAH CAFE.

CORNER OF FIRST CLASS RECEPTION ROOM, R.M.S. "OLYMPIC."

Drawing Room, R.M.S. "Olympic."

◀◀ Postcards
Views of the *Olympic* were often used to represent her younger sister as well. (*Clockwise from top*): The first-class verandah café on the *Olympic*; the drawing room as it existed on the *Olympic*; the first-class reception room on the *Olympic*. (All George Behe collection.)

▲ Postcard
The first-class lounge as it existed on the *Olympic* and *Titanic*. A visitor to the latter vessel on sailing day later wrote, 'Then there was the large library and reading room, with its many shelves of calf-bound, gilt-edged volumes, and its comfortable armchairs, and the spacious writing-room, where some of the passengers were already writing letters on the ship's notepaper, headed "RMS *Titanic*, at Sea." I wonder whether those letters will ever be read?' (George Behe collection.)

FIRST CLASS CAFÉ PARISIEN. R. M. S. "OLYMPIC".

▲ Postcard
A Café Parisien existed on the *Olympic* and *Titanic*. 'I have now explored the ship, except the Turkish bath and the swimming-bath,' first-class passenger Henry Julian wrote to his wife. 'The Parisian cafe is quite a novelty and looks very real. I do not know to what extent it is patronized, but it will, no doubt, become popular amongst rich Americans.' (George Behe collection.)

FRAGMENTS OF HISTORY: THE SHIP

FIRST CLASS RESTAURANT, R.M.S. "OLYMPIC".

"OLYMPIC" RESTAURANT.

▶ **Postcards**
Two views of the first-class restaurant on the *Olympic*. (George Behe collection.)

FIRST CLASS DINING SALOON, R.M.S. "OLYMPIC".

FIRST CLASS BEDROOM OF PARLOUR, R.M.S. "OLYMPIC".

◀◀ Postcards

(*Clockwise from top*): The *Olympic*'s first-class dining saloon. 'There was the great first-class dining-room, where I think they said six hundred persons could dine at once,' a visitor to the *Titanic* wrote later. 'There were scores of tables for parties of from two to eight. I recall a sensation of thick pile carpets, spotless napery, glittering silver, and countless flowers; and you entered by wide door-ways from the large crush room where the guests gathered before and after meals.' (George Behe collection.); A Tudor-style first-class cabin as it existed on both the *Olympic* and *Titanic*. (George Behe collection.); A first-class bedroom on the *Olympic*. 'The *Titanic* was assuredly the last thing in comfort and luxury afloat,' a visitor noted before the ship left Southampton. 'The "cabins" of the first class were not cabins but rooms furnished in lavish richness by Maple's. There were Marie Antoinette bedrooms, bedrooms with lovely, old-fashioned English four-poster wooden bedsteads, with old rose canopies and valences, and bedrooms done in the old Dutch style. There were family suites with beautiful sitting-rooms and servants' quarters. One wondered how even in so huge a ship room could be found for so much.' (George Behe collection.)

▶ First-Class Deck Plan

Passengers who contemplated sailing on the *Titanic* were given deck plans by accommodating travel agents so that they could familiarise themselves with the ship's layout and choose which cabin they wished to book. This deck plan is the extremely rare 29 March 1912 version that was issued just twelve days before the commencement of the great vessel's maiden voyage. The lady who received this plan from Cook's Shipping Office decided not to sail on 10 April but kept the plan as a souvenir anyway. (George Behe collection.)

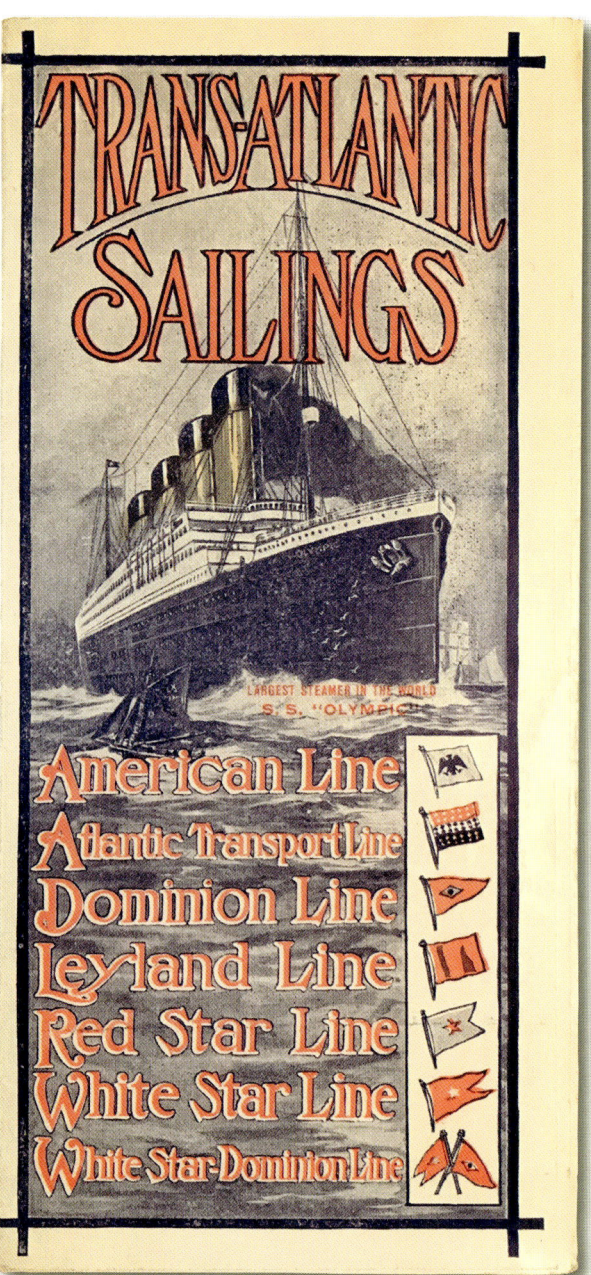

▲ **IMM Fold-Out Brochure**
For the 1912 season, the International Mercantile Marine Company supplied this large fold-out brochure for European travellers. It shows the various routes available and has advertising for companies under IMM. *Titanic* and *Olympic* are featured under White Star Line's section. (Mike Beatty collection.)

▲ **IMM Sailing Schedule**
Sailing schedules like this one helped prospective passengers choose a convenient time to make a transatlantic voyage on the *Olympic*, *Titanic* or other vessels of the White Star Line. (George Behe collection.)

▲ **Bournville Cocoa Card**
The White Star Line found creative ways to promote their new ships. These tiny cards advertising the *Titanic* and *Olympic* were issued by Cadbury's Chocolates. (Mike Beatty collection.)

◀ **Cadbury's Bournville Chocolate Tin**
The lid of this advertising tin bears a view of the ship painted by Montague B. Black. (Kalman Tanito collection.)

▼ **Passenger Booklet**
This booklet from an April 1912 voyage of the *Adriatic* features advertising for *Titanic*'s maiden voyage, which was to occur the following week. (Mike Beatty collection.)

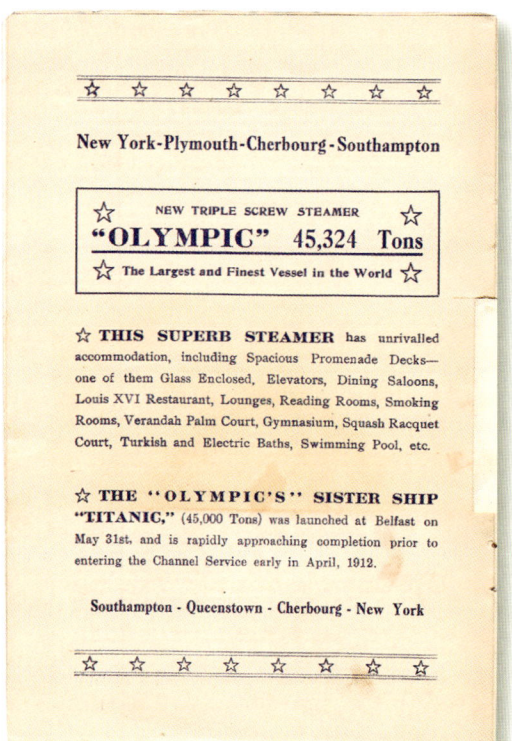

▶ **First- and Second-Class Accommodation Brochure**
The White Star Line went to great lengths to advertise both the *Titanic* and the *Olympic*. In this rare booklet, colourful artwork of the first- and second-class accommodations entice would-be travellers with the luxuries available on board. (Mike Beatty collection.)

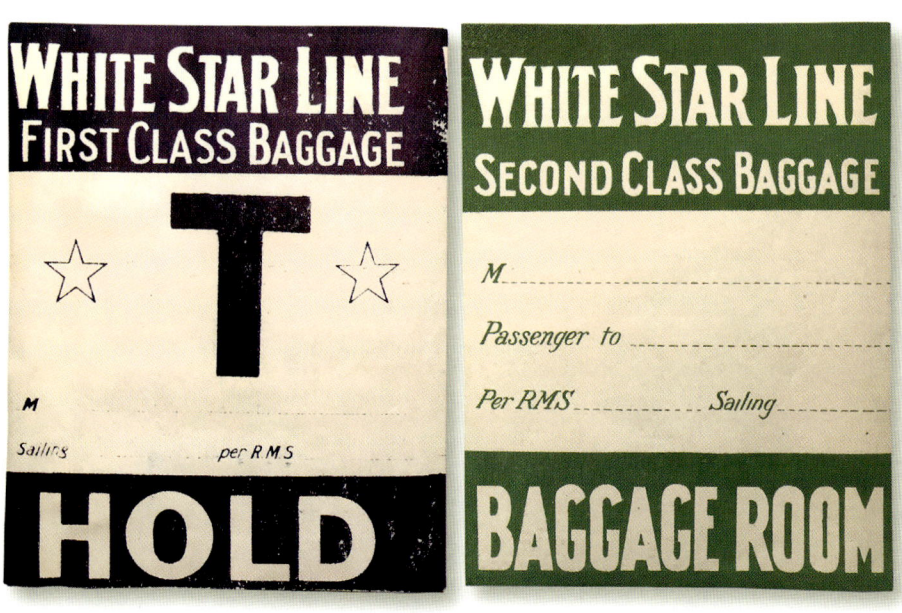

▲ ▶ ▼ **Luggage Stickers and Tags**
When a passenger's trunks and suitcases were brought to the White Star dock, stickers or cardboard tags were affixed to each piece of luggage to identify the owner and tell crewmen whether the item should be taken to the cargo hold or carried to the passenger's cabin for their use during the voyage. (George Behe collection.)

▶ ▶ **Beaver Felt Top Hat and Case**
This top hat (which was out of fashion by 1912) is representative of the type of headgear worn by first-class passengers during the era prior to the *Titanic*. This hat came with its own leather carrying case and is adorned with stickers from the finest hotels in Europe. It also boasts a faded circular White Star sticker, indicative that its owner travelled first class on a White Star Line vessel. (John Lamoreau collection.)

FRAGMENTS OF HISTORY: THE SHIP

◀ **Vinolia Otto Soap Advertisement**
This famous advertisement featuring the *Titanic* appeared in *The Illustrated London News* on 6 April 1912 – just four days before the vessel's maiden voyage. Passengers who sailed on the *Titanic* would find Vinolia Otto soap available for their use on board. (Mike Beatty collection.)

FRAGMENTS OF HISTORY: THE SHIP 67

▶▼ *The Green Book of Prophecies*
This 1912 book of predictions of the future has an eerie entry for the month of April. (Mike Beatty collection.)

MAIDEN VOYAGE

◀ Postcard

This unusual view of the *Olympic* leaving Southampton undoubtedly mirrors the view the residents would have had of the *Titanic*'s passengers as the big White Star liner pulled away from the dock to begin her maiden voyage. (George Behe collection.)

◀ Postcard

On 10 April 1912, tugs manoeuvred the *Titanic* away from the White Star dock to begin the first leg of her maiden voyage. 'Weather is very fine but a bit cold,' first-class passenger George Graham wrote in a letter to a friend. 'There is not a very big passenger list, about four hundred. It is a beautiful boat and I have a very fine state room on C deck right in the middle of the boat.' (George Behe collection.)

▲ Postcard
A view of the approaching *Titanic* by a photographer on board the docked vessel *Beacon Grange*. 'You do not notice anything of the movement of this ship, but the weather is very fine,' second-class passenger Henry Hodges wrote to a friend. (George Behe collection.)

▲ Postcard
Another photo of the *Titanic* from the *Beacon Grange*. 'I think the *Titanic* a most lovely boat + you hardly know you are moving,' second-class passenger Thomas Mudd wrote to his family. 'Will write more later. With love to all, Tom.' (George Behe collection.)

▲ Postcard
The *Titanic* proceeds on her way. 'I like my cabin very much it is like a bed-sitting room and rather large,' first-class passenger Adolphe Saalfeld jotted in a letter to his wife. 'I am the first man to write a letter on board.' (George Behe collection.)

◀ Postcard
Not long after this photograph was taken, the *Titanic* was steaming past the moored liner *New York* when the huge amount of water she was displacing snapped the smaller ship's mooring lines and pulled the helpless vessel out into the channel. (George Behe collection.)

▲ Postcard

It was only by the smallest of margins that the *New York* was prevented from colliding with the *Titanic*, but eventually the brand-new White Star liner was able to proceed on her way. 'They tell me that when we left Southampton Docks we drew the *New York* from her moorings, and nearly as possible ran into her,' second-class passenger Kate Buss wrote in a letter that evening. 'The tugs that should have pulled us out had to rush to her assistance and drag her away. I understand that the captain was a very interested spectator, but I saw nothing of it myself.' (George Behe collection.)

◀ **Postcard**
After her narrow scrape from the *New York*, the *Titanic* continued steaming down the Solent towards the open sea. 'After leaving at noon we had quite a little excitement,' Adolphe Saalfeld wrote to his wife, 'as the tremendous suction of our steamer made all the hausers of the SS *New York* snap as we passed her and she drifted on to our boat, a collision being only averted by our stopping and our tugs coming to the rescue of the *New York*. You will probably have read of the accounts in the papers.' (George Behe collection.)

◀ **Postcard**
The *Titanic* proceeds towards Cherbourg, her first port of call, to pick up additional passengers and mail. 'I had quite an appetite for luncheon,' Adolphe Saalfeld wrote to his wife, 'soup, fillet of plaice, a loin chop with cauliflower and fried potatoes, apple Manhattan & Roquefort cheese washed down with a large spaten beer iced, so you see I am not faring badly.' (George Behe collection.)

FRAGMENTS OF HISTORY: THE SHIP   73

▲ Postcard
In this rare pre-disaster postcard view, the *Titanic* reaches open water and begins to pick up speed. 'In the language of the poet, "This is a knock-out,"' first-class passenger Arthur Gee wrote in a letter that day. 'I have never seen anything so magnificent, even in a first-class hotel. I might be living in a palace. It is, indeed, an experience. We seem to be miles above the water, and there are certainly miles of promenade deck. The lobbies are so long that they appear to come to a point in the distance.' It's unknown why the postcard's publisher chose to illustrate Lord Nelson's flagship *Victory* along with the *Titanic*. (George Behe collection.)

▲ Passenger List
As the *Titanic*'s maiden voyage proceeded, passengers found themselves perusing passenger lists like this one to learn the identities of their fellow travellers. 'There are only 370 1st Class Passengers,' Adolphe Saalfeld wrote in a letter to his wife. 'So far the boat does not move and goes very steady. It is not nice to travel alone and leave you behind. I think you will have to come next time.' (George Behe collection.)

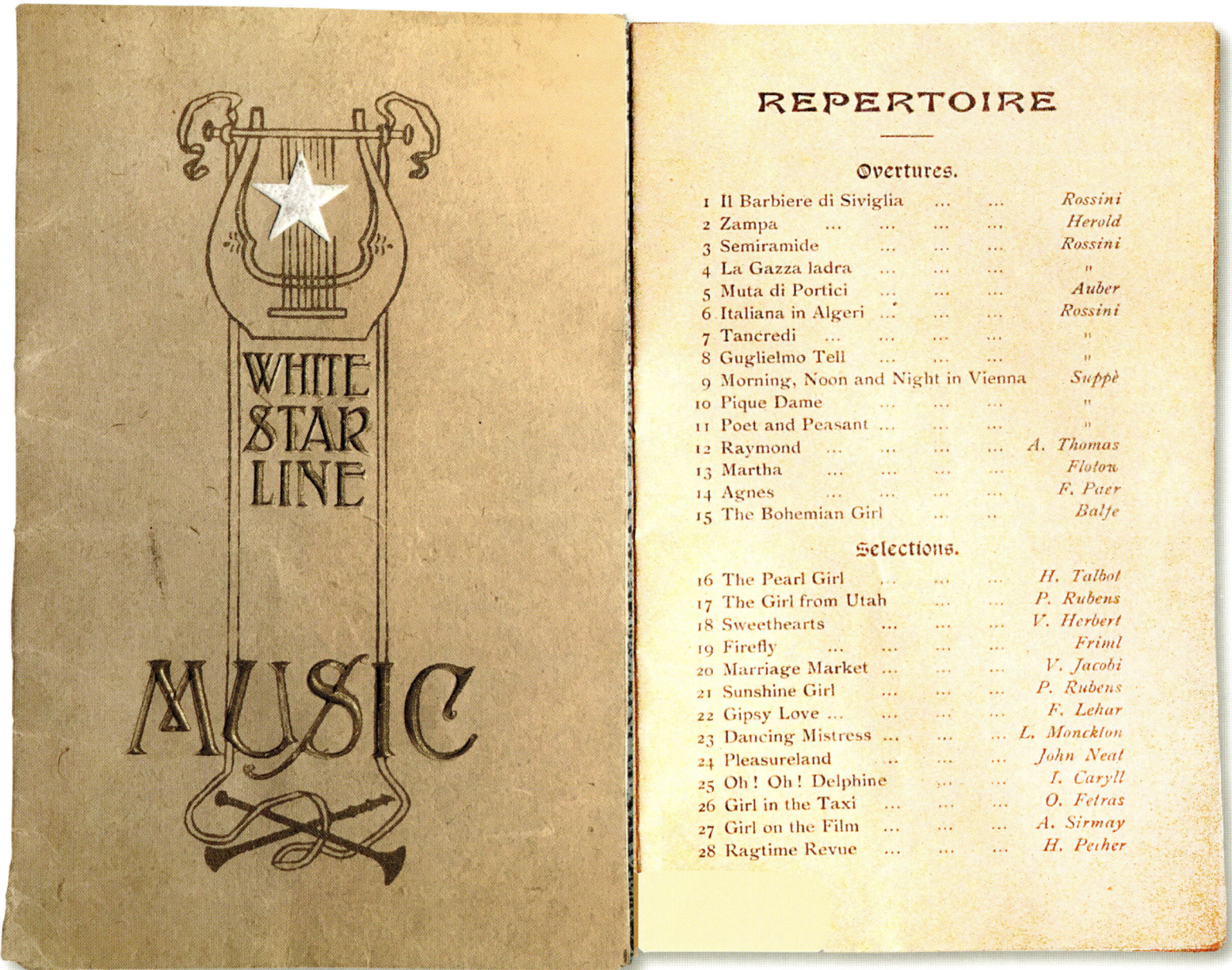

▲ **Music Booklet**
One popular pastime for passengers was listening to music played by the *Titanic*'s bandsmen, who were prepared to perform musical pieces that passengers could choose from small booklets like this one. 'There are two bands, one in the lounge and the other in the café,' Henry Julian wrote in a letter to his wife. On the night of 14–15 April, the *Titanic*'s eight bandsmen were destined to achieve undying fame for the gallant behaviour they displayed during the ship's evacuation. (George Behe collection.)

FRAGMENTS OF HISTORY: THE SHIP   75

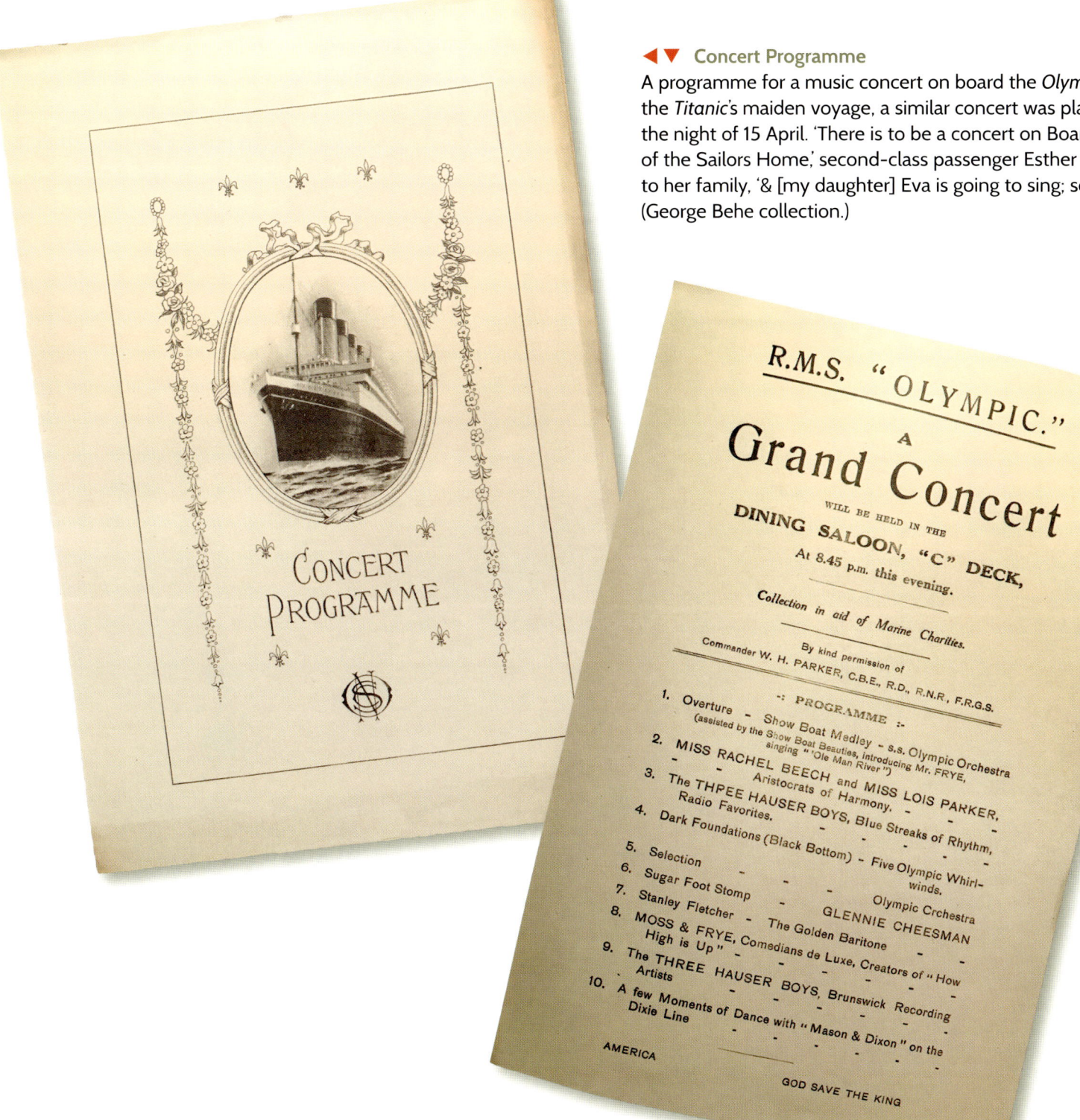

◀▼ **Concert Programme**

A programme for a music concert on board the *Olympic*. During the *Titanic*'s maiden voyage, a similar concert was planned for the night of 15 April. 'There is to be a concert on Board … in aid of the Sailors Home,' second-class passenger Esther Hart wrote to her family, '& [my daughter] Eva is going to sing; so am I.' (George Behe collection.)

◀ **First-Class Steamer Rug**
During the voyage, a first-class passenger would have relaxed in a deck chair with a plaid wool steamer rug like this one. (Mike Beatty collection.)

▶ **Third-Class Cabin Berth Blanket**
Amenities in third class on the *Olympic* and *Titanic* were better than second class in other ships of the time. Here a surviving original blanket, as used on board the two ships, gives an idea of sleeping conditions in a third-class berth. (Mike Beatty collection.)

▶ **White Star Envelope**
Another typical shipboard pastime was the writing of letters to friends and loved ones on shore. Ship's stationery embossed with the *Titanic*'s letterhead was available to passengers, who inserted their completed letters into White Star envelopes, like this one, before giving them to stewards for delivery to the ship's mail room. (George Behe collection.)

◀ **Olympic Lettercard**
This perforated lettercard is engraved with the *Olympic*'s name. Identical cards bearing the name *Titanic* were used on board the younger White Star liner. (George Behe collection.)

▶ **Turkish Bath Ticket**
The *Titanic* and *Olympic* were both noted for their luxurious accommodations, and a Turkish bath was one of the special amenities offered to first-class passengers. This ticket to the *Titanic*'s Turkish bath came from a page of perforated tickets that was removed from the vessel in Southampton before she began her maiden voyage. 'The cold is better and voice is coming back,' an ailing Alfred Rowe would write to his wife on 11 April. 'I took a lovely Turkish bath yesterday and that did me good …' (George Behe collection.)

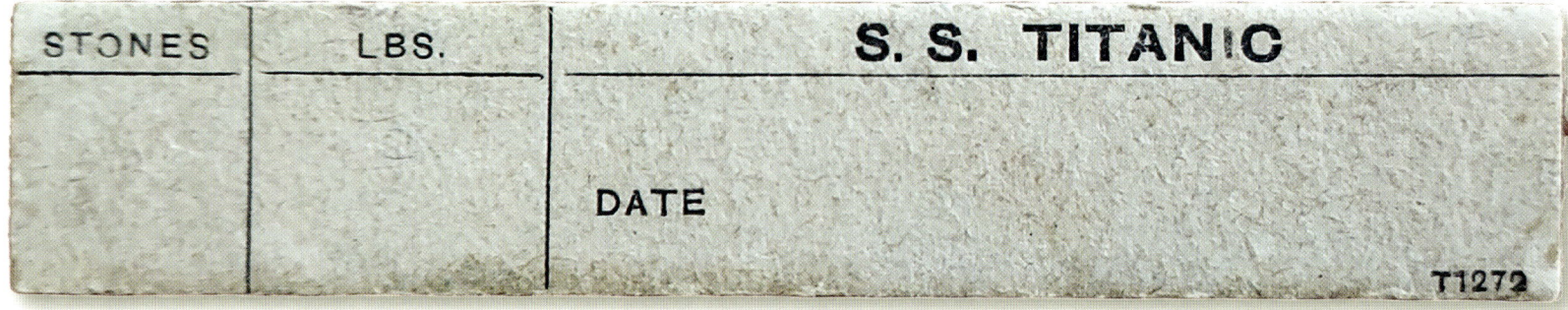

▲ **Weighing Ticket**
A rare ticket from the personal weighing machine located within the Turkish baths. The machine, manufactured by Henry Pooley & Son of Birmingham, recorded the weight of the passenger and printed it on the ticket as a souvenir. The document originally belonged to William Cole, an employee of Henry Pooley & Son, who installed the weighing machine on board and subsequently kept the ticket as a memento. (Trevor Powell collection.)

▲ **Bookmark**
A pre-sinking ivory souvenir bookmark for the *Titanic*. Incidentally, the illustration appears to be that of Cunard's *Lusitania/Mauretania*. (Trevor Powell collection.)

▲ Playing Cards

Another pastime for the *Titanic*'s passengers was playing poker and bridge whist in the ship's first-class smoking room. Three professional gamblers on the *Titanic* spent their time trying to fleece fellow passengers at the card table, and they undoubtedly used White Star playing cards identical to the ones shown here. 'After luncheon [on 14 April] my husband asked me if I would sit in, if necessary, in a poker game,' Renée Harris wrote in later years. 'He explained that in a previous "session" one of the players had been under suspicion, and rather than bar him from the game, this time it would be simpler to let him see that the table was filled. It turned out that I was called on to be the eighth "man." When the suspected person was pointed out to me I thought he was a minister of the gospel, he looked so virtuous.' (Mike Beatty collection.)

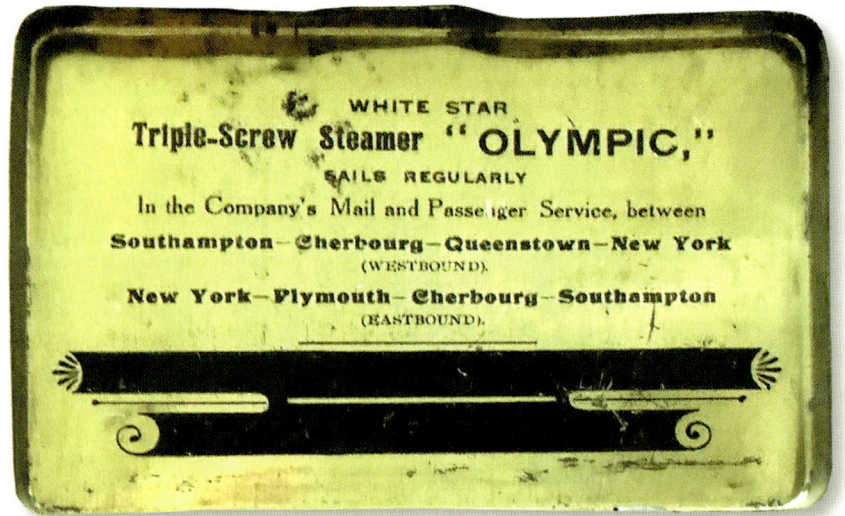

▲ **Cigarette Tin**
These tins advertising the *Olympic* and *Titanic* were sold as souvenirs in the barber shops on board both vessels. The inside covers of these tins advertised the *Titanic* as still being under construction, but after the disaster her name was blacked out so that the White Star Line could sell its remaining stock of tins without being encumbered with unusable merchandise. (George Behe collection.)

FRAGMENTS OF HISTORY: THE SHIP 81

▶ **Dining Room Chair**

A *c.* 1911 mahogany chair from the second-class dining saloon of the *Olympic*. The cast-iron swivel base allowed the chair to be turned when bolted to the floor. The chair, which features an Art Nouveau design, was hand-crafted by the skilled carpenters at Harland & Wolff. (Trevor Powell collection.)

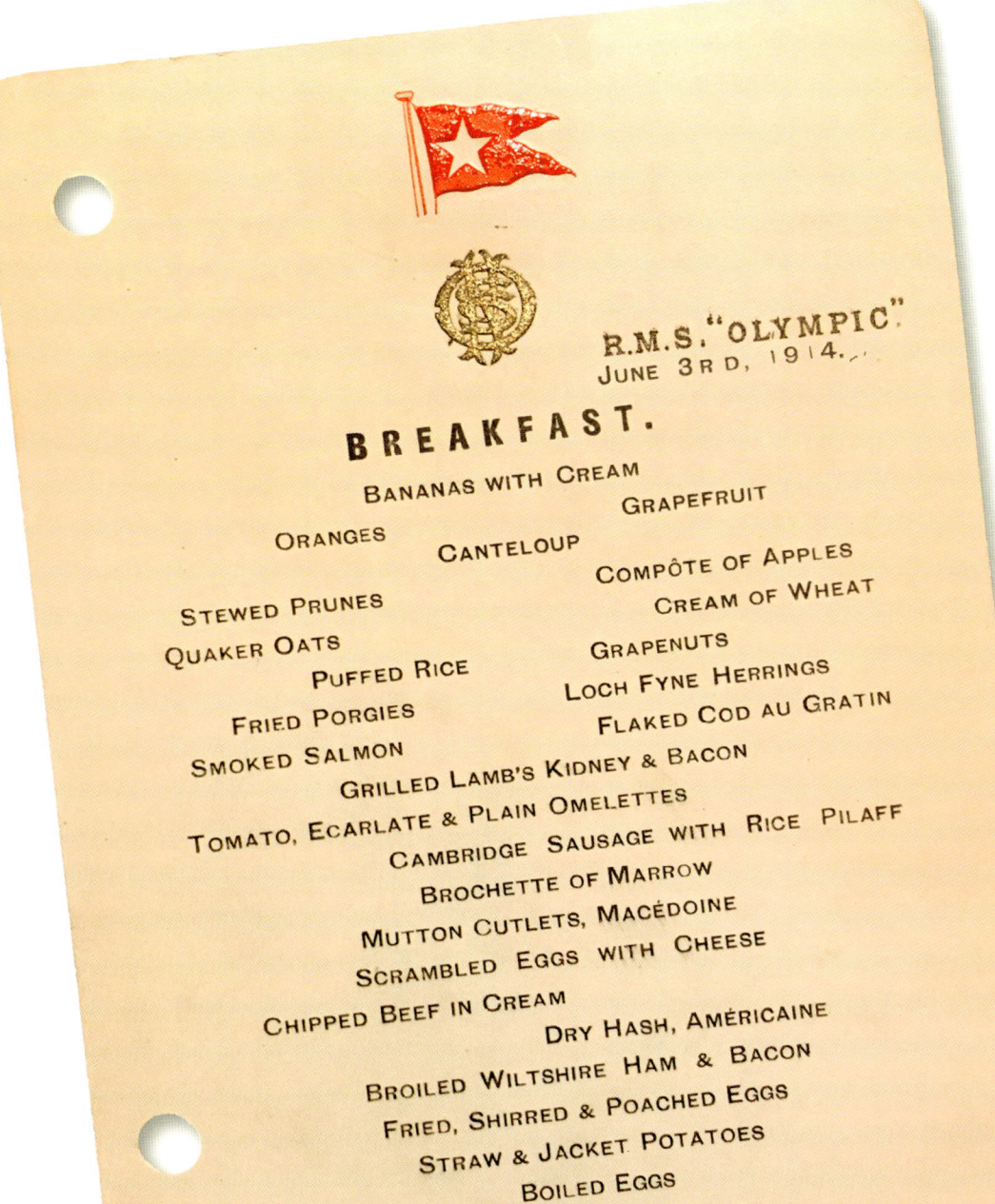

▲▶ **Menus**

Shown here are examples of the *Olympic*'s first-, second- and third-class menus. The menu styles on the *Titanic* were identical to these patterns. 'I had a table with two old ladies for lunch, & another table tonight for dinner but no people, bad luck,' dining room steward Harry Bristow wrote to his wife on the evening of 10 April. 'I hope I shall get a permanent table tomorrow, when most of the passengers come aboard or else it will mean nil for me as tips are concerned.' (George Behe collection.)

FRAGMENTS OF HISTORY: THE SHIP

Triple Screw Steamer "OLYMPIC."

2ND CLASS
June 19, 1912

DINNER

Consomme Julienne

Broiled Halibut, Anchois

Calves' Feet en Poulette
Roast Haunch of Mutton, Currant Jelly
Casserole of Chicken

Carrots & Turnips   Green Peas
Browned & Boiled Potatoes

Cold—Cumberland Ham
Salad

Diplomate Pudding
Almond Blancmange
Cocoanut Cakes
Horton's Ice Cream—Wafers
Cheese   Biscuits
Apples   Oranges   Water Melon
Tea   Coffee

THIRD CLASS
R.M.S. "OLYMPIC."
July 30, 1912

BREAKFAST.
Boiled Cerealine & Milk
Creamed Salt Cod
Vegetable Stew
Fresh Bread & Butter
Marmalade   Swedish Bread
Tea   Coffee

DINNER.
Pea Soup
Corned Brisket of Beef & Cabbage
Boiled Potatoes
Fresh Bread
Sago Pudding
Oranges

TEA.
Lancashire Hot Pot
Fresh Bread & Butter
Stewed Apples & Rice
Tea

SUPPER
Gruel   Cheese

Any complaint respecting the Food supplied, want of attention or incivility, should be at once reported to the Purser or Chief Steward. For purposes of identification, each Steward wears a numbered badge on the arm.

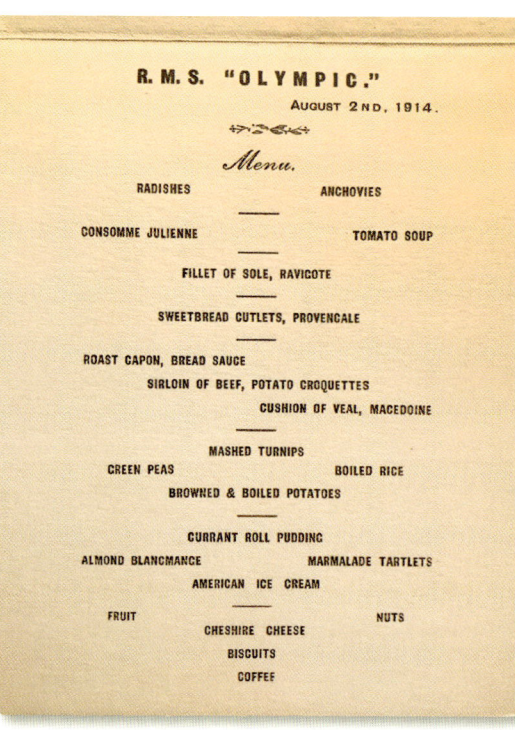

◀ **Menu**
Another style of menu used on board the *Olympic* and *Titanic*. (George Behe collection.)

▼ **First-Class White Star China**
A 'Gothic Arch' pattern dinner plate and a 'Wisteria' pattern salad plate. (Mike Beatty collection.)

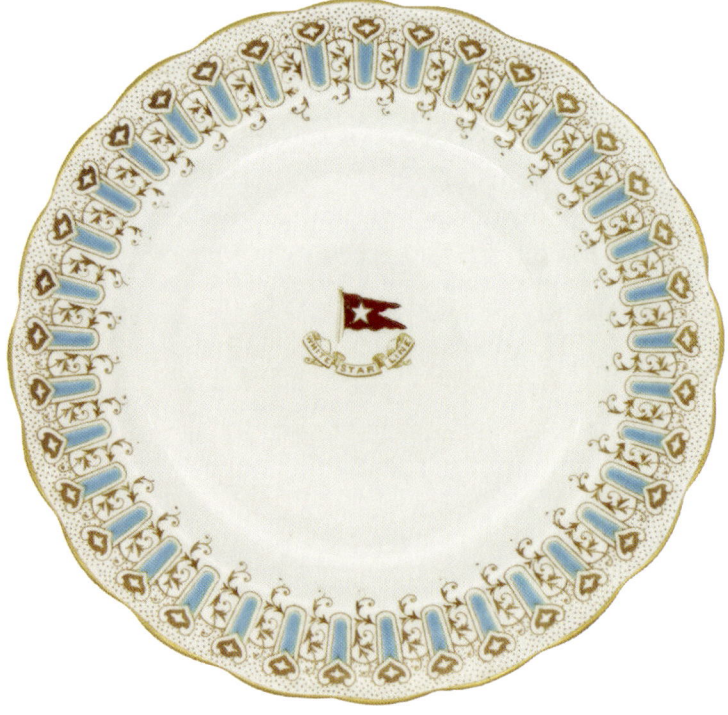

FRAGMENTS OF HISTORY: THE SHIP   85

▲ **Second-Class White Star China**
A very rare set of a second-class bouillon cup and saucer. This pair were reported to have been used aboard the *Olympic*. This pattern, known as 'Blue Delft', was used by the White Star Line beginning in the late nineteenth century up until the First World War. (Don Lynch collection.)

▲ **Second-Class White Star China**
A White Star Line second-class asparagus dish, made by Stonier & Co. Ltd, Liverpool. (Trevor Powell collection.)

▲ **Second-Class White Star China**
These examples of second-class china were manufactured by Stonier & Co. Ltd, from Liverpool. (Trevor Powell collection.)

▼ **Third-Class White Star China**
Third-class passengers were served meals on simple white dishes emblazoned with the White Star burgee. Here is a soup plate and cup as used on board. These examples were recovered from the sunken White Star vessel *Arabic*. (Mike Beatty collection.)

▲ **White Star Deck Service**
(Top): A boullion plate meant to be used with a two-handled boullion cup on deck in first class. *(Bottom):* A 'Wisteria' butter pat (left) is shown next to the equivalent used for deck service. (Mike Beatty collection.)

▶ **White Star Hors D'Oeuvres Dish**
Very few of these rare hors d'oeuvres dishes have survived to the present day. (Mike Beatty collection.)

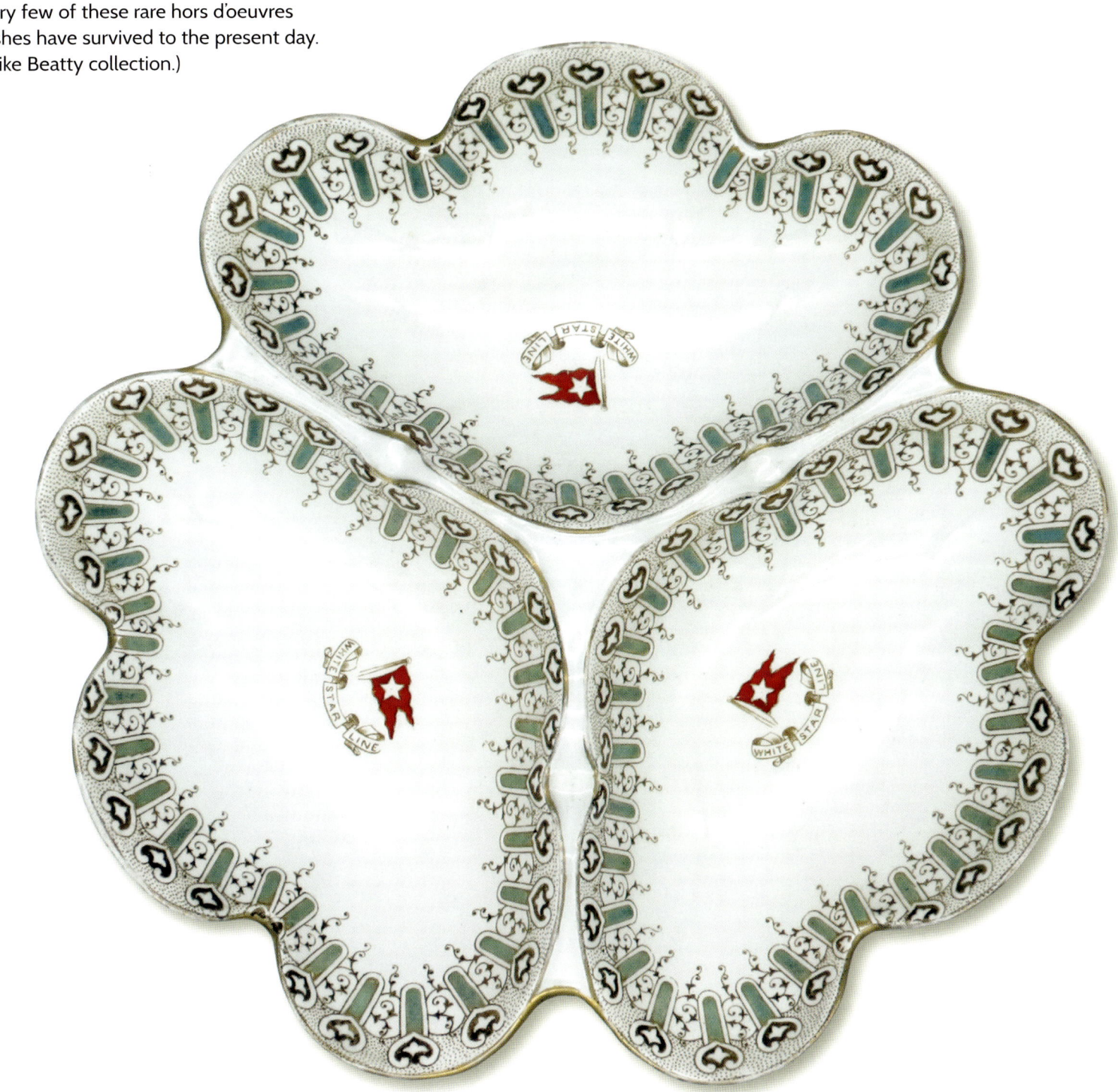

▶ **White Star China**
Possibly a prototype set of china for the White Star Line, in a very uncommon, but beautiful, pattern. (Kalman Tanito collection.)

▶ **China**
The famous R4332 Spode pattern, Rd.No.580303, is the rarest *Titanic* china of all. The 'Greek Key' design is printed in gold on a cobalt blue band, with gold lines and raised dots. (Kalman Tanito collection.)

FRAGMENTS OF HISTORY: THE SHIP   89

▲ Napkin Ring
The *Titanic* offered all the best amenities to its first-class passengers, and among them were the small napkin rings that were used in the ship's dining room. (George Behe collection.)

▶ White Star Line Carafe
This carafe was recovered from the wreck of the RMS *Republic* in 1981. Identical examples were recovered from the *Titanic*'s wreck site. (Trevor Powell collection.)

▲ **Postcard**
A French postcard utilises a photograph of the *Olympic* to depict the *Titanic* as she arrived at Cherbourg at about 6:30 p.m. on 10 April 1912. After picking up additional passengers and mail, the *Titanic* turned her bows toward Queenstown, Ireland, prior to repeating the same process the following day. 'We reached [Cherbourg] in nice time and took on board quite a number of passengers,' Harland & Wolff ship designer Thomas Andrews wrote to his wife. 'The two little tenders looked well; you will remember we built them about a year ago. We expect to arrive at Queenstown about 10.30 a.m. tomorrow. The weather is fine and everything shaping for a good voyage. I have a seat at the Doctor's table.' (George Behe collection.)

"Titanic, triple Screw, 45,000 tons, Largest Steamer in the World."

▶ **Postcard**
After steaming through the night, the *Titanic* arrived at Queenstown shortly before noon on 11 April to pick up more passengers and mail. 'Have had a fine run round to Queenstown,' first-class passenger Richard Smith wrote on a postcard that was taken off the ship at that Irish port. 'Just leaving for the land of Stars and Stripes. Hope you are all well at home.' The *Titanic* then turned her bows westward towards New York and headed out onto the vast expanse of the North Atlantic. (George Behe collection.)

▲ **Postcard**
As the *Titanic* steamed away from Queenstown, steerage passenger Eugene Daly stood on the vessel's poop deck and played his uilleann pipes as a poignant farewell to his homeland. That same afternoon the clouds that had been blocking the sun slowly began to scatter, and the wind gradually died away as darkness approached. (George Behe collection.)

◀ **Postcard**
On the morning of 12 April the *Titanic*'s passengers sat down to a hearty breakfast in the ship's dining rooms. 'I remember being childishly pleased on finding strawberries on my breakfast table,' first-class passenger Lucy Duff Gordon recalled later. '"Fancy strawberries in April, and in mid-ocean. The whole thing is positively uncanny," I kept saying to my husband. "Why, you would think you were at the Ritz."' (George Behe collection.)

▶ **Postcard**
On 12 April, steerage passenger Jacob Johansson wrote the following entry in his diary: 'Beautiful weather no wind we have an excellent accommodation everything is clean and tidy big promenade deck and light and fresh roe.' (George Behe collection.)

▶ **Postcard**
At 5:46 p.m. on 12 April, the *Titanic* received a wireless message intended for the eyes of Captain Edward J. Smith: 'To Capt. *Titanic*. My position 7 p.m. GMT lat. 49.28 long. 26.28 W. dense fog since this night crossed thick ice field lat. 44.58 long. 50.40 "Paris" saw another ice field and two icebergs lat. 45.20 long. 45.09 "Paris" saw a derelict lat. 40.56 long. 68.38 "Paris" please give me your position best regards and bon voyage. Caussin.' (George Behe collection.)

◀ **Postcard**

On 12 April, first-class passenger Arthur Gee began writing a lengthy letter to his wife: 'Excuse the pencil, but I want to keep picking this up from time to time & adding in things as they occur. I shall get a sample menu card of all these meals on board to give you an idea of what our stomachs have to perform. I sent a big correspondence off from Queenstown, hope you have received it all. So far the weather has been charming. On our big floating town we hardly feel any motion at all. I cannot say how it will behave if we get it rough, but the gale seems to have blown itself out.' (George Behe collection.)

◀ **Postcard**

A pre-maiden voyage postcard advertising the *Olympic* and *Titanic*. This card has been autographed by Winnifred Quick Van Tongerloo, a second-class passenger who survived the disaster. 'The first few days at sea were lovely,' Winnifred's mother, Jane Quick, once wrote. 'The water was calm. The air was brisk. It gave one an appetite, a zest for the games of the daytime. There was dancing, and shows, and deck games. Everyone was sociable. A sort of holiday spirit prevailed.' (George Behe collection.)

FRAGMENTS OF HISTORY: THE SHIP   95

▶ **Postcard**
Many of the *Titanic* passengers were impressed with the vessel's state-of-the-art technology. 'There is every possible convenience on ship,' second-class passenger Kate Buss wrote. 'For instance, when you go into the lavatories and bedrooms they are in darkness until you close the door, which is connected with a clip to the electric light, you open the door and the light goes off.' (George Behe collection.)

▶ **Postcard**
On the afternoon of 13 April, passenger Elizabeth Lines was seated in the *Titanic*'s first-class reception room when she overheard White Star chairman J. Bruce Ismay discussing the ship's speed with Captain Smith. 'My attention was arrested by hearing the day's run discussed,' she testified later, '… and I heard Mr Ismay … give the length of the run, and I heard him say, "Well, we did better today than we did yesterday … we will make a better run tomorrow. Things are working smoothly, the machinery is bearing the test, the boilers are working well." They went on discussing it, and then I heard him make the statement: "We will beat the *Olympic* and get into New York on Tuesday."' (George Behe collection.)

◀ **Return Voyage**

A flyer for the *Titanic*'s intended return voyage from New York on 20 April 1912. This notice was published by the White Star Line's New York office on 13 April, while the ship was already at sea on her first voyage. (Kalman Tanito collection.)

▶▶ **Postcards**

(*Clockwise from top left*): On this card the *Olympic* stands in for her younger sister *Titanic*, where second-class passenger Kate Buss concluded the latest instalment of her letter on the evening of 13 April. 'Goodbye,' she wrote. 'Hope you won't worry. I've got a [customs] declaration sheet to fill in. Don't know however I shall do it: it's a beastly nuisance. Tomorrow is Sunday. Hope it will keep fine.' (George Behe collection.); After encountering a bit of rain early on the morning of 14 April, the *Titanic* continued her way westward under sunny skies. 'Well, the sailors say we have had a wonderful passage up to now,' second-class passenger Esther Hart wrote in a letter to her mother. 'There has been no tempest, but God knows what it must be [like] when there is one. This mighty expanse of water, no land in sight & the ship rolling from side to side is very wonderful tho [sic] they say this ship does not roll on account of its size. Any how it rolls enough for me. I shall never forget it.' (George Behe collection.); During the afternoon of 14 April the weather turned colder, and by evening few passengers felt like braving the icy, speed-induced wind that swept across the ship's open decks. 'It's nice weather but awfully windy & cold,' Esther Hart wrote. 'They say we may get into New York Tuesday night, but we are really due early on Wednesday morning. Shall write as soon as we get there. This letter won't leave the ship but will remain & come back to England where she is due again on the 26th. Where you see the letter all of a screw is when she rolls & shakes my arm. I am sending you on a menu to show you how we live.' (George Behe collection.)

FRAGMENTS OF HISTORY: THE SHIP   97

T.S.S. TITANIC

▶ **Postcard**

Another pre-maiden voyage card on which *Olympic* substitutes for the *Titanic*. 'We have met some nice people on Board ... & so it has been nice so far,' Esther Hart wrote in her 14 April letter, 'but oh the long long days & nights – it's the longest week I ever spent in my life. I must close now with all our fondest love to all of you, From your loving Esther.' (George Behe collection.)

◀ **Postcard**

Although her official arrival time at New York's Ambrose Light was slated for 5 a.m. on Wednesday, the speed *Titanic* achieved during her maiden voyage would have resulted in her arriving at the Light late Tuesday night, which her elder sister *Olympic* was achieving with great regularity. Meanwhile, as darkness fell on the evening of 14 April, *Titanic* sped onward towards her destiny. (George Behe collection.)

DISASTER

▶ **Postcard**

Beginning at about 11:15 on the night of 14 April, *Titanic*'s lookouts are reported to have sighted several nearby icebergs passing at a safe distance on either side of the vessel. This post-disaster postcard illustrates what these sightings may have looked like as the ship passed through a known, documented belt of icebergs that lay on the eastern fringe of the main icefield. (George Behe collection.)

◀ Postcard

At 11:40 p.m. on 14 April, *Titanic* brushed against the side of an iceberg that extended slightly higher than her boat deck. 'Sunday night I was awakened by a jolt, and the engines stopped,' second-class passenger Nellie Becker wrote later. 'I heard people running about and went out into the hall to see what the matter was. A man said, "Nothing," so I went back into the room and lay down again.' (George Behe collection.)

◀ Postcard

After the collision, the *Titanic* listed slightly and began sinking by the bow. After completing his inspection trip below decks, Thomas Andrews hurried up to the bridge and told Captain Smith that the *Titanic* had less than ninety minutes remaining afloat. 'Then I heard people rushing again,' Nellie Becker continued in her written account, 'and I looked out the door and there stood our cabin Steward. I asked: "What is the trouble?" He said, "Tie on your life-belt and go on deck at once." I said: "Have we time to dress?" He said: "No, Madame, you have time for nothing."' (George Behe collection.)

FRAGMENTS OF HISTORY: THE SHIP   101

▶ **Postcard**

This photograph (originally published in *The Illustrated London News*) shows a Marconi wireless operator jotting down a Morse code message just like the *Titanic*'s senior operator Jack Phillips did after the collision. 'Mr. Phillips told me that apparently we had struck something,' junior operator Harold Bride recalled, 'as previous to my turning out [of bed] he had felt the ship tremble and stop, and expressed an opinion that we should have to return to Belfast. I took over the telephone from him, and he was preparing to retire when Capt. Smith entered the cabin and told us to get assistance immediately. Mr. Phillips resumed the phones, after asking the captain if he should use the regulation distress call "C Q D." The captain said "Yes," and Mr. Phillips started in with "C Q D," having obtained the latitude and longitude of the *Titanic*'. (George Behe collection.)

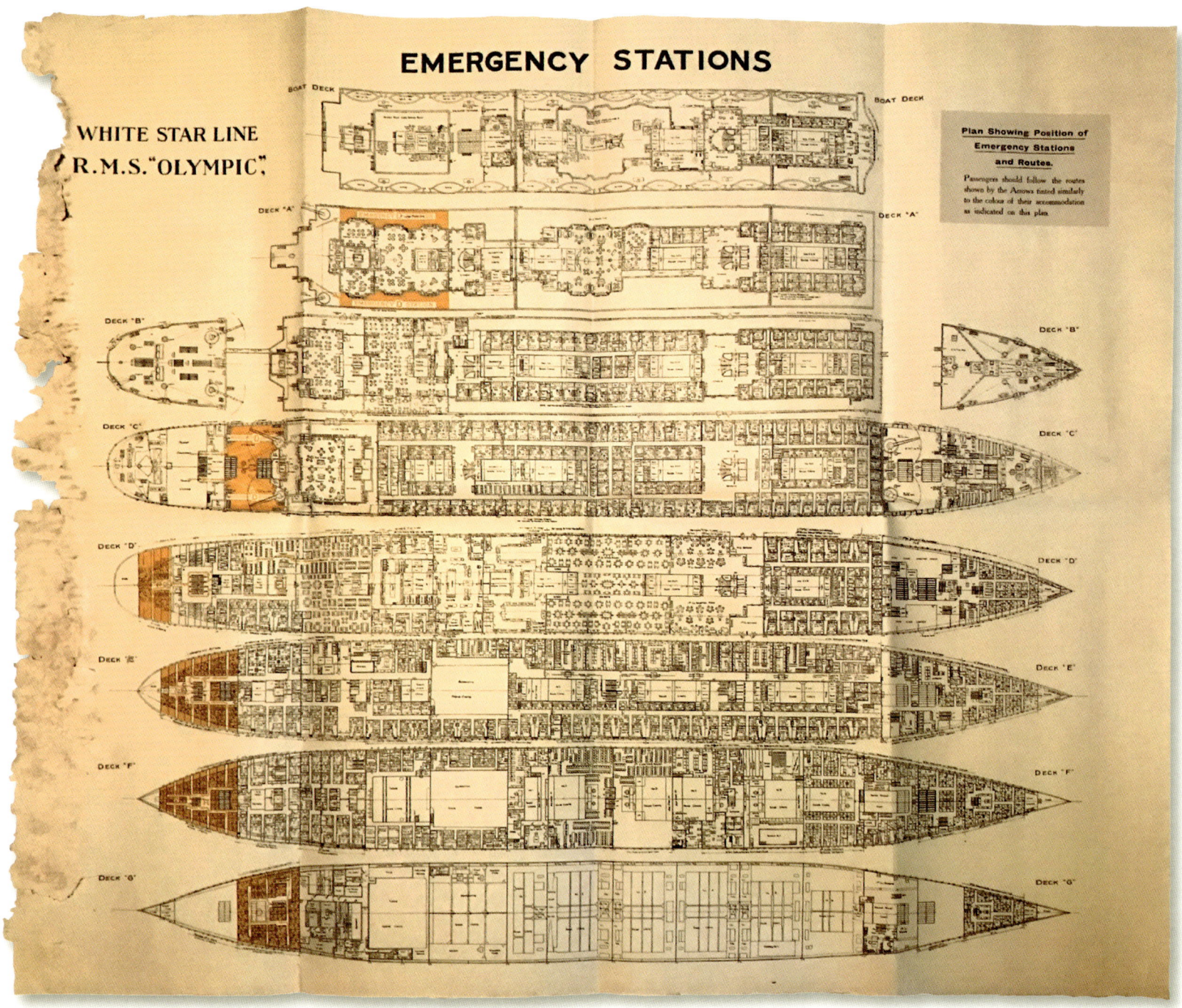

▲ **Emergency Stations Deck Plan**
One of a series of emergency station deck plans that was used on board the *Olympic*. This plan dates from the *Olympic*'s post-1912 career, and it's unknown if similar plans were posted for the benefit of the *Titanic*'s passengers during her voyage. (George Behe collection.)

▲ Postcard

As the *Titanic*'s bows continued to settle lower and lower into the water, her officers ordered passengers into the lifeboats and began lowering those half-filled boats down to the ocean's surface. 'We finally got in the fourth boat,' first-class passenger Kornelia Andrews wrote later. 'There was no panic, but no discipline. We had one sailor, and then any man who could row was allowed to get in and so a Chinese [Japanese] and an Armenian got in, saying they could row, but they could not, so Gretchen [Longley] assisted the sailor on one side and two or three women on the other, until her hands were frozen stiff. We pulled away from the ship so that we would not be drawn in with the suction should the ship go down. All women and children in the boats, and these ignorant men put in our boats while those splendid Americans stood and were not allowed to come with us.' (George Behe collection.)

◀ ▲ **Postcards**

(*Clockwise from left*): Many heart-rending scenes were enacted on the night of 14–15 April as wives and children were parted from husbands and fathers on the *Titanic*'s slanting decks. 'Oh, it would have broken your heart to have seen them standing there so bravely and waving farewell to their wives and daughters,' Kornelia Andrews remembered. (George Behe collection.); Sea conditions that night were much calmer than the ones depicted in this postcard illustration or else it's possible no lifeboats would have survived the night. 'It was now about 1 a.m.; a beautiful night, with no moon and so not very light,' second-class passenger Lawrence Beesley wrote later. 'The sea was as calm as a pond, just a gentle heave as the boat dipped up and down in the swell; an ideal night, except for the bitter cold, for anyone who had to be out in the middle of the Atlantic Ocean in an open boat, and if ever there was a time when such a night was needed, surely it was now with hundreds of people, mostly women and children, afloat hundreds of miles from land.' (George Behe collection.); As the *Titanic*'s forward decks submerged beneath the ocean's surface, efforts were redoubled to launch all remaining lifeboats while there was still time to do so. 'Presently,' Lawrence Beesley remembered, 'about 2 a.m., as near as I can remember, we observed her settling very rapidly, with the bows and the bridge completely under water and concluded it was now only a question of minutes before she went down, and so it proved.' (George Behe collection.)

▶ **Sheet Music**

During the evacuation, the *Titanic*'s eight bandsmen gathered inside the boat deck's first-class entrance and played cheerful music to allay the fears of alarmed passengers. 'At this moment the band was playing "Alexander's Ragtime Band",' first-class passenger May Futrelle remembered. 'What a scene that was – the men of the first cabin, many of them still in evening clothes, with drawn, set faces, who but a few short minutes ago were in command of all the things of the world, now found themselves face to face with the specter of death. Did they flinch? Not one. They were not the kind of men who quailed in the face of danger. Oh, their courage was superb!' (George Behe collection.)

▼ **Postcard**

At around 2 a.m., the *Titanic*'s bandsmen stepped out onto the open boat deck to perform their last few pieces of music and, from her lifeboat, Marie Jerwan watched and listened to what happened next. 'Little by little the lights disappeared one after another, until we could see only a black mass,' Mrs Jerwan wrote later. 'The bow was already submerged. We still heard the musicians of the ship playing the beautiful hymn: "Nearer My God to Thee", to which we joined in with all our heart. What heroism to stay that way at their post to give courage to those who were going to die, in playing this song, so beautiful and so solemn.' None of the *Titanic*'s eight bandsmen survived the sinking. (George Behe collection.)

▲ **Postcard**
As the *Titanic*'s bridge and forward boat deck began to submerge, Duane Williams and his son Richard found themselves standing nearby with the *Titanic*'s forward funnel towering high above them. 'Still remembering "suction" I yelled to my father "quick – jump,"' Richard remembered in later years. 'He started towards me just as I saw one of the four great funnels come crashing down on top of him. Just for one instant I stood there transfixed … It was probably but a fraction of a second, then I jumped to the rail; the water was about two and a half to three decks down. I climbed over the rail and jumped clear.' (George Behe collection.)

FRAGMENTS OF HISTORY: THE SHIP

▲ **Postcard**
As the *Titanic*'s bridge submerged, the unlaunched lifeboat collapsible B was swept off the boat deck and capsized in the swirling water. Colonel Archibald Gracie can be seen here after he swam away from the ship and encountered the upturned boat. 'I threw my right leg over the boat astraddle, pulling myself aboard, with a friendly lift to my foot given by someone astern as I assumed a reclining position with them on the bottom of the capsized boat,' Colonel Gracie wrote later. Two years after the *Titanic* disaster, an enterprising artist changed the 1912 illustration by painting out the *Titanic*'s forward funnels to use the altered version to commemorate the 1914 loss of the two-funnelled *Empress of Ireland*. (George Behe collection.)

Captain Smith, saving a child while the ship went down - April 15th 1912

▲ **Postcard**
An artist's impression of Captain Edward J. Smith handing a dying child up to occupants of the overturned lifeboat collapsible B. Although this story is often regarded as apocryphal, survivor August Wennerström later wrote that he saw Smith pick up one of steerage passenger Alma Pålsson's four children right before the *Titanic*'s bridge began to submerge. Neither Captain Smith, Mrs Pålsson nor her four children survived the night. (George Behe collection.)

▶ **Postcard**

The *Titanic*'s stern continued to rise out of the sea until it finally reached an angle of about 30 degrees. Many passengers sheltering on the stern were unable to find handholds and found themselves sliding or tumbling down the ship's slanting decks towards the hungry sea. 'She slowly tilted straight on end with the stern vertically upwards,' Lawrence Beesley remembered, 'and as she did, the lights in the cabins and saloons which had not yet flickered for a moment since we left, died out, came on again for a single flash, and finally went out altogether. At the same time the machinery roared down through the vessel with a rattle and a groaning that could be heard for miles, the weirdest sound surely that could be heard in the middle of the ocean, a thousand miles away from land.' (George Behe collection.)

◀ **Postcard**

The immense weight of *Titanic*'s rising stern finally caused it to break away from the submerged bow section and fall back into the sea. 'We had just got beyond the reach of the suction,' survivor Kate Buss wrote. 'It was a grand sight at sea, and with every light on she was a picture. She parted right in halves, the forward part went down first, and the aft seemed to stand upright. There was a terrific explosion; the cries of the souls on board were awful to hear, but there was absolutely nothing to be seen.' Unlike the view depicted on this postcard, the *Titanic*'s bow section did not re-emerge above the ocean's surface, but instead broke completely free from the now-fallen horizontal stern and began its long fall towards the ocean floor. (George Behe collection.)

◀ **Postcard**

After floating in its normal horizontal position for a brief time, the *Titanic*'s broken stern section suddenly rose again towards the vertical and then plunged straight down into the sea. 'I shall never forget the sight of that beautiful boat as she went down, the orchestra playing to the last, the lights burning until they were extinguished by the waves,' Dr Alice Leader wrote. 'It sounds so unreal, like a scene on the stage.' Only a couple of men who jumped from the plunging stern section lived to tell the tale. (George Behe collection.)

RESCUE

▲ Postcard
The Cunard liner *Carpathia* was roughly 58 miles from the *Titanic* when her Marconi operator received the stricken White Star liner's distress call. 'Then the first officer was blurting out the facts [to me],' Captain Arthur Rostron remembered, 'and you may be sure I was very soon wide awake, with thoughts for nothing but doing all that was in the ship's power to render the aid called for.' Captain Rostron immediately ordered that his vessel be turned out of its normal course, and the rescue ship began steaming north-west towards the disaster site. (George Behe collection.)

▲ **Postcard**
The *Carpathia* reached the disaster site at around 4 a.m. on 15 April. Survivor Nellie Becker wrote, 'After drifting about till daylight, all but frozen, and not knowing whether the wireless messages had been received, we saw a large boat in the distance and then the rockets from her to tell us she had come to our rescue.' (George Behe collection.)

▶ *Carpathia* Postcard

The *Carpathia* immediately became the focus of the survivors, who began to row their lifeboats towards the rescue ship. 'As we neared the *Carpathia*,' Lawrence Beesley remembered, 'we saw in the dawning light what we thought was a full rigged schooner standing up near her and presently behind her another, all sails set, and we said: "They are the fisher boats from the Newfoundland banks and have seen the steamer lying to and are standing by to help." But in another five minutes the light showed pink on them and we saw they were icebergs towering many feet in the air, huge glistening masses, deadly white, still, and peaked in a way that had easily suggested a schooner.

We glanced round the horizon, and there were others wherever the eye could reach. The steamer we had to reach was surrounded by them and we had to make a detour to reach her, for between her and us lay another huge berg. We rowed up to the *Carpathia* about 4:30 a.m., and were hoisted or climbed up the ship's sides with very grateful hearts.' (George Behe collection.)

▶ Postcard

This card shows one of *Titanic*'s lifeboats (collapsible D) approaching the *Carpathia*. 'Never can anyone know how we felt when we knew we were saved,' wrote survivor Nellie Becker, who was saved in another boat, 'and we went with light hearts to [the *Carpathia*], thinking that very few were lost. We were so numb with cold when we came to the boat that we could not get out. The Officers of the *Carpathia* had to come and get us out. The children they lifted in gunny sacks. Then they took us and tied us into a sort of swing and lifted us up. When they got us up, the Stewards lifted us up and carried us into the saloon, gave us brandy and wrapped us up in blankets.' (George Behe collection.)

◀▲ **Louis Ogden Rescue Photos**
On the morning of 15 April 1912, *Carpathia* passengers Louis Ogden and his wife, Augusta, woke up to find their ship stopped, surrounded by ice, and lifeboats rowing towards them. Louis grabbed his camera and captured history in the making. *Titanic* survivor Archibald Gracie used Louis Ogden's photos for his book *The Truth About the Titanic* and likely made these two enlargements to give to the Ogdens. These photos show lifeboat 16 rowing up alongside the side of the *Carpathia*. (Mike Beatty collection.)

FRAGMENTS OF HISTORY: THE SHIP

▶ **Postcard**

A photograph of *Titanic*'s rescued survivors on board the *Carpathia*. 'You simply cannot imagine the sadness here,' survivor Nellie Becker wrote. 'Only two Stewardesses out of fourteen were saved; almost all the Stewards gone; only one or two left. About a dozen little orphaned children are here. Of course nobody saved a thing. Many have not a cent. Some tried to take handbags, but the officers just took them and threw them into the sea. On the *Carpathia* we stay at day in the saloon and at night in the Officers' quarters, for the boat is full up.' (George Behe collection.)

▶ **Postcard**

This card depicts a typical burial at sea of a passenger whose life ended unexpectedly during a ship's mid-passage. Four similar burials of *Titanic* victims took place from the decks of the *Carpathia* on the afternoon of 15 April 1912. 'Some of the rescued passengers died afterward on the *Carpathia*, and were buried at sea,' survivor Vera Dick wrote. 'There were none of the usual formalities. It was desired that no attention be directed to the occurrences. The rescued passengers were frantic enough already.' (George Behe collection.)

▼▶ **Iceberg**

A 15 April 1912 marconigram sent by Captain W. Wood of the *Etonian* to the *Olympic* through the *Asian*. It wouldn't command much attention, but in that area, he also photographed an iceberg that he thought sank the *Titanic*. For some reason, he put the year in the inscription as 1914, and in 2020 Henry Aldridge and Son sold another example of the photo dated 1913. (Kalman Tanito collection.)

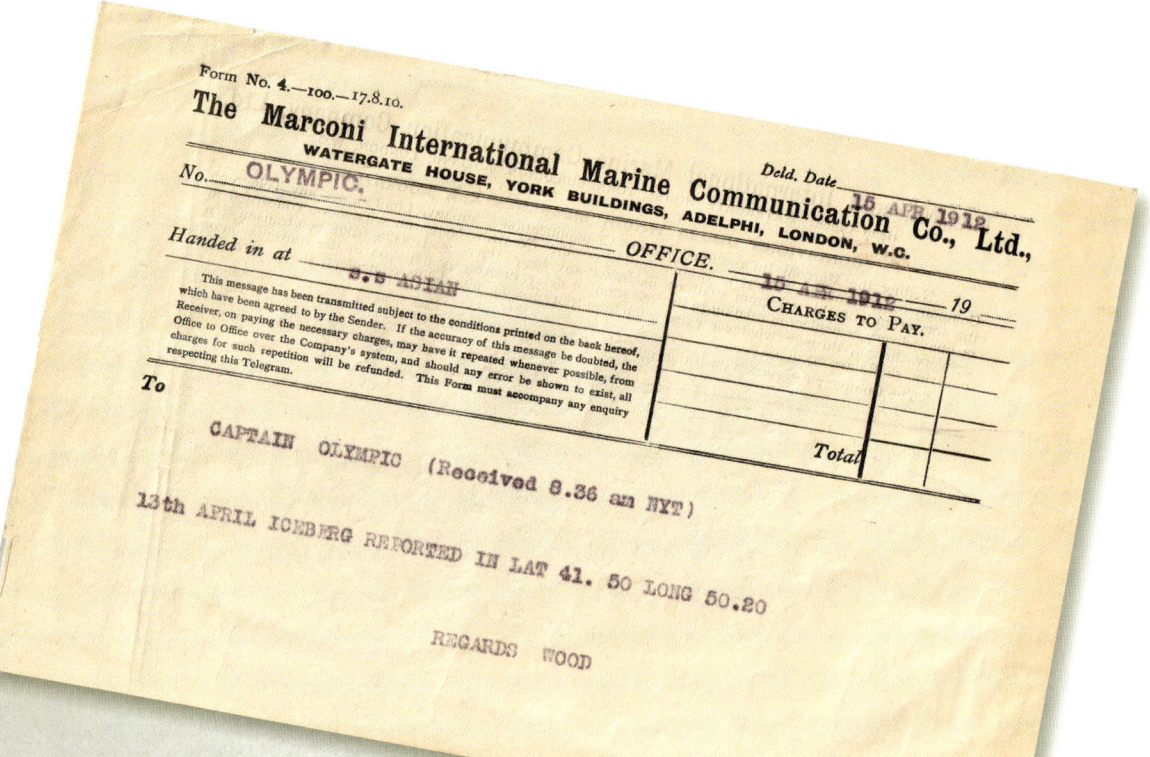

FRAGMENTS OF HISTORY: THE SHIP   117

▶ *Carpathia* Postcard
'One poor fellow on our deck I saw some hours after we came aboard,' survivor Kate Buss wrote on 16 April. 'He is the only one I knew to speak to that is saved; quite a lad. I was so delighted that I felt I must kiss him. He had an awful time, and is aching with bruises from floating wreckage. He had six hours in the water before he was picked up, and he told me there were 50 in the boat who died, but he didn't know who they were, it was so dark.' (George Behe collection.)

▶ *Carpathia* Postcard
'They have looked after us wonderfully well,' Kate Buss wrote on 16 April, 'and I am fortunate in sharing a berth with a nurse, given up by a gentleman, when I preferred to sleep on the deck than going right below steerage in the sailors' berths: I felt I should smother if I did. I have had neuralgia in my head fearfully, but then I cried so much when I knew we had lost our friends, and it's simply awful to see the distressed widows and distracted parents. Never as long as I live shall I forget it nor the brave souls who, I know, have perished.' (George Behe collection.)

◀ *Carpathia* Postcard

'I have not seen Mr & Mrs Allison or Loraine so I suppose they have gone under,' survivor Amelia Brown wrote on 17 April, 'but there is just the chance they may have been picked up by another ship. I'm not going to worry about it, and they have several friends on board, and then there are the partners of the firm. We have been offered a home until something is found for us.' (George Behe collection.)

◀ *Carpathia* Postcard

Carpathia was still headed towards New York on 17 April. 'I lie awake at night expecting to be called up at every little sound,' a nervous Kate Buss wrote on that date. 'I hear that there has been a lot of fog for days, and we hear noises on this vessel that we never heard on the *Titanic*. We have some awfully kind-hearted folks on board. Yesterday we stated our case before a committee of ladies and gentlemen, who intend to get as much as possible from the White Star Company.' (George Behe collection.)

▶ *Carpathia* Postcard

'My nerves are all to pieces, and we are in a dense fog at present,' survivor Gladys Cherry wrote in a letter on 17 April. 'We ought to get to New York tomorrow ... it's all been too ghastly, I can't blame anyone and they say the Captain shot himself; but there seemed no discipline on the boat, as there ought to have been an officer in each boat – there were only 16 lifeboats and 2 canvas ones for 2,600 passengers, there are 490 passengers and 210 crew saved 700 altogether out of nearly 3,000, isn't it awful? Why did we go at that pace when they knew we were near ice bergs?' Even though only 2,208 people were on board the *Titanic*, Miss Cherry's horror at the death rate was very real; two out of three people who sailed on the *Titanic* died with her. (George Behe collection.)

◀ Teletype

A 1912 teletype message detailing messages from the *Carpathia* as it sailed for New York City with the *Titanic*'s survivors. The message is from a scrapbook created by William A. Turner, who worked at the New York Stock Exchange in 1912 and continued working there until the late 1930s. Turner was present when members of the Stock Exchange met the *Carpathia* at the dock and gave '$20,000 in cash to be distributed among *Titanic* sufferers.' (John Lamoreau collection.)

◂ *Carpathia* Postcard
'Here on the *Carpathia* we see, my dear, such sad things,' Kornelia Andrews wrote in a letter to her niece on 18 April. 'Nearly every other woman is weeping for her husband or child and it is all past description. They are kind, so kind. I have had three staterooms offered me, but there are others who need them so much more, and so we slept on the floor in the library with 25 or 30 others, babies crying because both parents were lost, and others because their mothers were gone. And to think it was all so unnecessary.' (George Behe collection.)

◂ *Carpathia* Postcard
'We are supposed to get into New York tonight, but we are still in fog,' survivor Gladys Cherry wrote in a letter on 18 April. 'Under ordinary circumstances they would not land us at night, but perhaps now they will, if you could only imagine how we long for land – This water all round is terrible, and one's nerves now seem worse than on that dreadful night. The Doctor gave me a little Bromide last night and I slept a little better, but one wakens terrified, which is very silly, as we have nothing to grumble at in comparison with the poor widows.' (George Behe collection.)

◀ *Carpathia* Postcard
On the evening of 18 April the *Carpathia* arrived in New York, where *Titanic*'s survivors were finally put ashore. 'At last we are here safe and sound,' survivor Gladys Cherry wrote in a later letter. 'Now that it is all over one feels full of gratitude and thankfulness for the deliverance from the dangers we have been brought through … It all seems like an ugly dream … We got into the dock last night at 9.30, after a dreadful time coming up the river, with all the newspaper tugs that wanted to put Pressmen on board, but of course, our captain would allow no one on board but the Pilot. All these people meeting the boat stood under their names, all the officials standing in lines to keep back the crowds; motors by the million were waiting.' (George Behe collection.)

▶ Postcard
This card shows *Titanic* lifeboats being unloaded from the *Carpathia* after she arrived in New York on the evening of 18 April 1912. Many of these boats had already been stripped of *Titanic*'s metal nameplates, numbers, and other accoutrements by souvenir hunters on board the rescue ship. (George Behe collection.)

▲ Silk *Carpathia* Postcards
Three extremely rare woven-in-silk *Carpathia* postcards. (George Behe collection.)

▶ **Cunard Bulletin**
A bulletin listing the various wireless messages received by the *Carpathia* during an earlier voyage. (George Behe collection.)

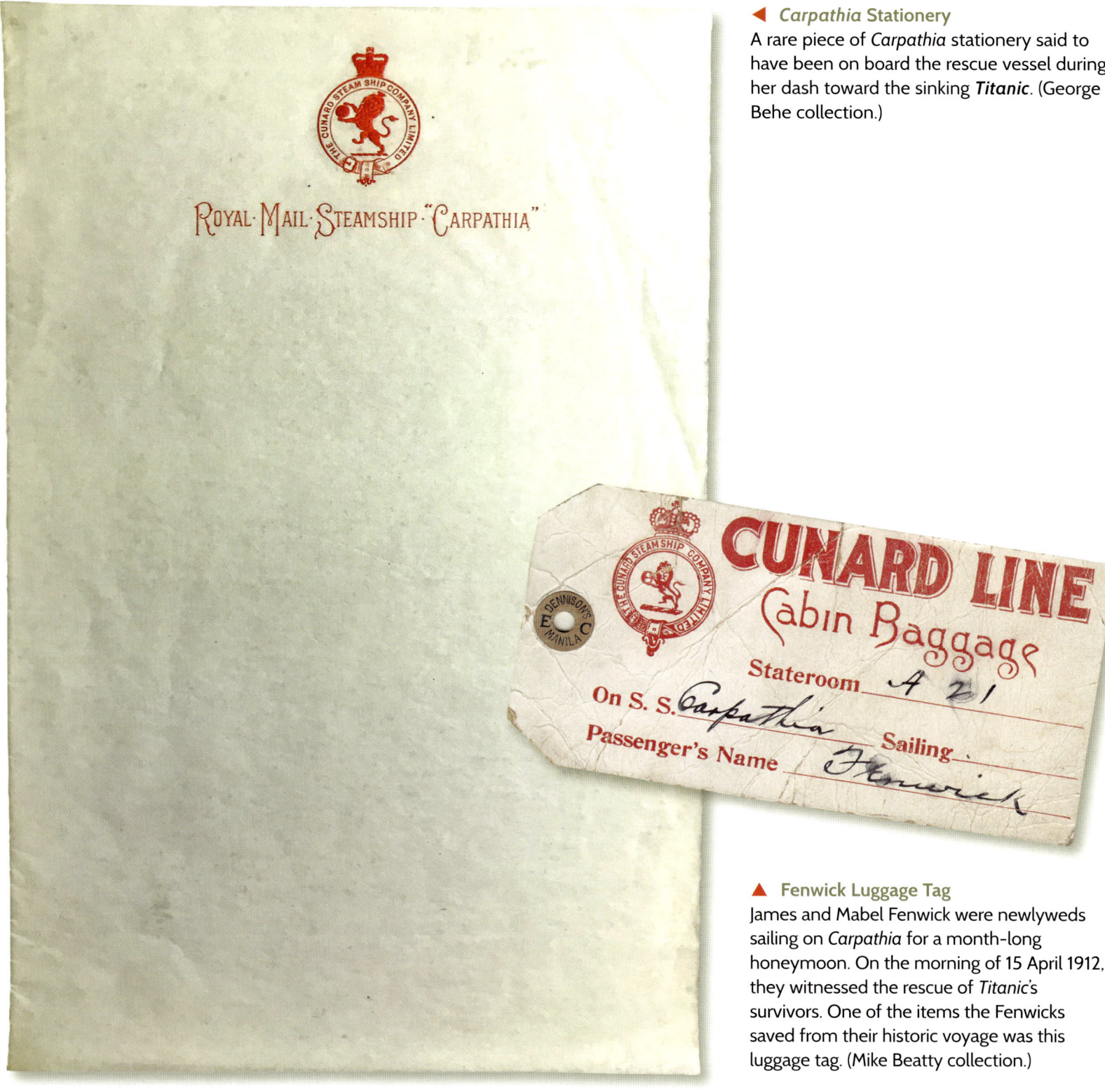

◀ *Carpathia* Stationery
A rare piece of *Carpathia* stationery said to have been on board the rescue vessel during her dash toward the sinking **Titanic**. (George Behe collection.)

▲ Fenwick Luggage Tag
James and Mabel Fenwick were newlyweds sailing on *Carpathia* for a month-long honeymoon. On the morning of 15 April 1912, they witnessed the rescue of *Titanic*'s survivors. One of the items the Fenwicks saved from their historic voyage was this luggage tag. (Mike Beatty collection.)

▶ **Dance Programme**

On 25 April 1912 – just ten days after the sinking – a dance was held in the *Carpathia*'s first-class dining saloon after the rescue ship resumed her voyage to the Mediterranean. This programme was saved by Mabel Fenwick, who documented the rescue operation through her camera lens just ten days prior. (Trevor Powell collection.)

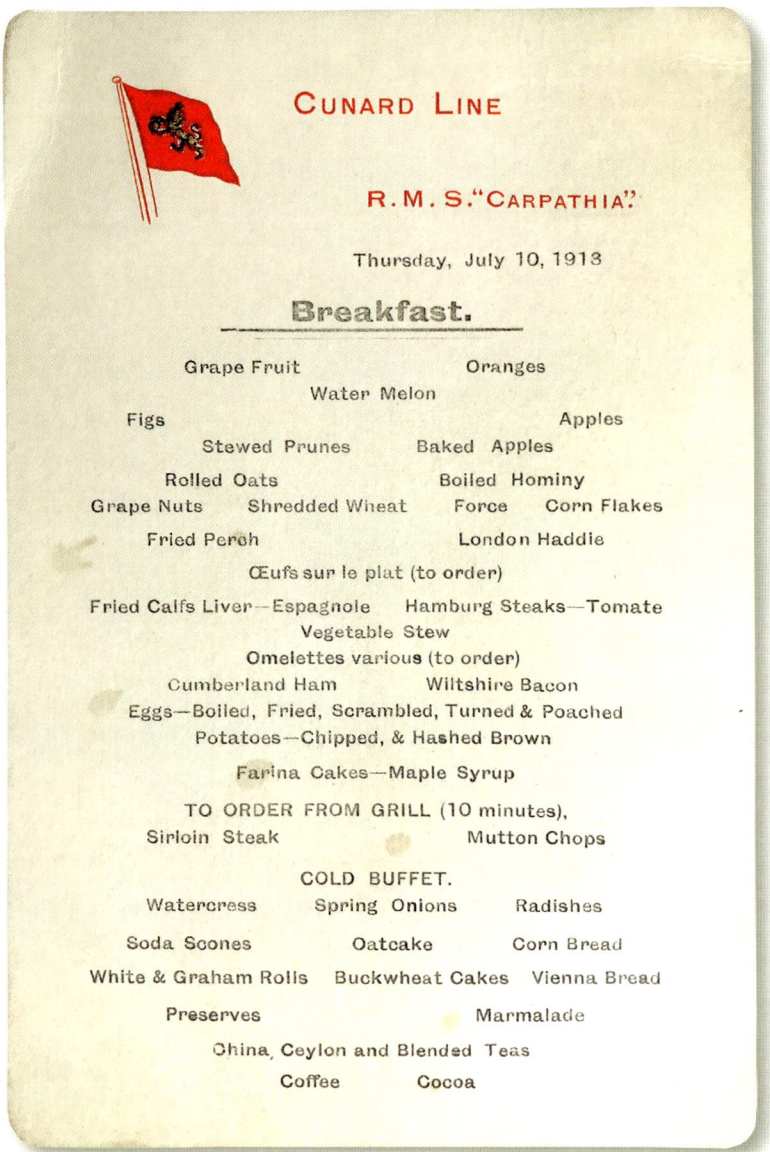

▲ *Carpathia* Menu
Louis and Augusta Ogden, who were on board the *Carpathia* during the *Titanic* rescue, sailed on that vessel again later in 1912. They saved a menu from that trip. (Mike Beatty collection.)

▲ *Carpathia* Menu
Another example of the rescue ship's menus. (George Behe collection.)

▶▼ *Carpathia* Voyage Photo Album
Six weeks after *Carpathia* made her historic rescue, another couple sailed on board the vessel, snapping photos on deck where many dramatic scenes had taken place just a few weeks before. Captain Rostron was captured in one photo, and the couple also pasted in a photo of the *Carpathia* bought on board. (Mike Beatty collection.)

Carpathia Voyage Photo Album
Shown here are a few additional photographs that were snapped on the *Carpathia* by the aforementioned couple which were later fastened into their photo album. (Mike Beatty collection.)

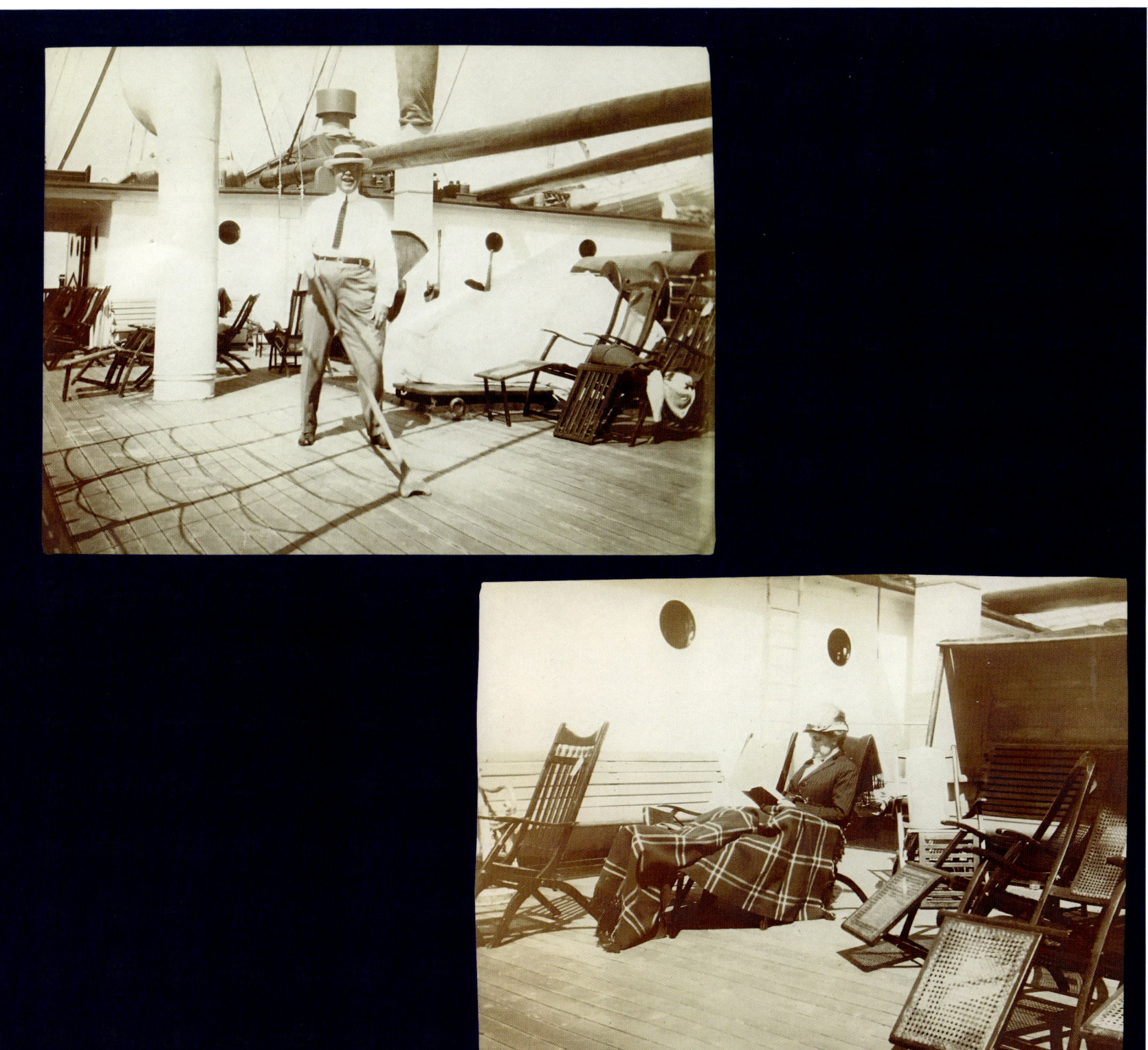

▲ **Bronze *Carpathia* Medal**
During the months after the *Titanic* disaster, survivor Margaret Brown ('The Unsinkable Molly Brown') commissioned medals to be struck for presentation to Captain Arthur Rostron and the crew of the *Carpathia* for their heroic actions during the rescue. This bronze medal was awarded to one of the vessel's ordinary crewmen. (George Behe collection.)

▶ **Captain Rostron Spicer Medal**
It's been lost to time who commissioned the making of this medal, but soon after the disaster artist Theodore Spicer-Simpson cast and distributed a limited number of these bronze medallions honouring Captain Arthur Rostron. This example, still in its original box, was owned by *Titanic* survivor Helen Churchill Candee. (Mike Beatty collection.)

FRAGMENTS OF HISTORY: THE SHIP 131

▲ Cunard Schedule
Inside this late April 1912 Cunard schedule are the changes in the *Carpathia*'s sailing schedule that were caused by her returning to New York with the *Titanic*'s survivors, which postponed her 11 April voyage. (Mike Beatty collection.)

▲ Souvenir Spoon
A sterling silver souvenir spoon from the *Carpathia*, hallmarked 1912. (Trevor Powell collection.)

◀ **Souvenir Brooch**
A silver and enamelled souvenir brooch from the *Carpathia*. Examples such as this would have been available for purchase in the ship's barber shop. (Trevor Powell collection.)

▶ ***Carpathia* Tiles**
On 17 July 1918, the rescue ship *Carpathia* met its demise during the dark days of the First World War when she was struck by three torpedoes and sunk off the coast of Ireland. Eighty-nine years later, a team of divers recovered this section of floor from the wreck. It's likely this tile felt the footsteps of many *Titanic* survivors for four days in April 1912. (Mike Beatty collection.)

FRAGMENTS OF HISTORY: THE SHIP   133

◀ *Carpathia* Chronometer Ring
This ring once held the chronometer – a crucial instrument in navigation – that would have been enclosed in a wooden box in the chart room. It was recovered from the wreck in 2007. (Mike Beatty collection.)

▶ *Carpathia* Light Fixture
This light fixture salvaged from *Carpathia* was used in the various public areas on board the rescue ship. (Mike Beatty collection.)

Form No. 4.—100.—17.8.10.

The Marconi International Marine Communication Co., Ltd.,
WATERGATE HOUSE, YORK BUILDINGS, ADELPHI, LONDON, W.C.

Deld. Date 15 APR 1912

No. OLYMPIC.   OFFICE. 15 APR 1912   19

Handed in at CARPATHIA

This message has been transmitted subject to the conditions printed on the back hereof, which have been agreed to by the Sender. If the accuracy of this message be doubted, the Receiver, on paying the necessary charges, may have it repeated whenever possible, from Office to Office over the Company's system, and should any error be shown to exist, all charges for such repetition will be refunded. This Form must accompany any enquiry respecting this Telegram.

CHARGES TO PAY.

Total

To

COMMANDER OLYMPIC   RECEIVED 5.20 pm NYT

DO YOU THINK IT IS ADVISABLE TITANIC'S PASSENGERS SEE OLYMPIC.
PERSONALLY I SAY NOT. ROSTRON.

▲ *Olympic* Marconigram
There was a lot of wireless communication between the *Carpathia* and the *Titanic*'s sister *Olympic* during the day of the rescue. This is one such message, received by the *Olympic*, that was used as evidence during the inquiries that followed. (Mike Beatty collection.)

FRAGMENTS OF HISTORY: THE SHIP

▶▼ *Olympic–Parisian* Marconigrams

As the *Olympic* races towards the *Titanic*, her wireless is busy trying to find information from other ships. Here are five Marconigram messages between the *Olympic* and the *Parisian* on the morning of the sinking. (Mike Beatty collection.)

Marconigram 1 (Olympic to Captain Parisian, sent 12.25 pm EST, 15 APR 1912):

MANY THANKS FOR MESSAGE CAN WE STEER TO 41.22 N 50.14 W FROM WESTWARD AND THEN NORTH TO TITANIC FAIRLY FREE FROM ICE WE ARE DUE THERE AT MIDNIGHT SHOULD APPRECIATE TITANIC CORRECT POSITION IF YOU CAN GIVE IT ME

Haddock

Marconigram 2 (S.S. Parisian to Olympic, received 9.30 am EST, 15 APR 1912):

PARISIAN SENT MESSAGES TO TITANIC AT EIGHT THIRTY LAST NIGHT AND HEARD TITANIC SENDING TRAFFIC TO CAPE RACE JUST BEFORE PARISIAN OPERATOR WENT TO BED AT ELEVEN FIFTEEN SHIPS TIME CALIFORNIAN WAS ABOUT FIFTY MILES ASTERN OF PARISIAN. HEARD FOLLOWING THIS

P.T.O

Marconigram 3 (Parisian to Olympic contd.):

PARISIAN TO OLYMPIC CONTD.

MORNING AT SIX OCLOCK:- ACCORDING TO INFORMATION PICKED UP CARPATHIA HAS PICKED UP TWENTY BOATS WITH PASSENGERS. BALTIC IS RETURNING TO GIVE ASSISTANCE AS REGARDS TITANIC I HAVE HEARD NOTHING. DONT KNOW IF SHE IS SUNK.

Marconigram 4 (Parisian to Captain Olympic, received 12.50 pm NYT, 15 APR 1912):

Safe from field ICE TO 41.22 50.14 AS THE ICE WAS YESTERDAY YOU WOULD NEED TO STEER FROM THAT POSITION ABOUT NE AND N. TO ABOUT LAT 41.43 AND 50.00 THEN APPROACH HIS POSITION FROM THE WESTWARD STEERING ABOUT WNW MY KNOWLEDGE OF THE TITANICS POSITION AT MIDNIGHT WAS DERIVED FROM YOUR OWN MESSAGE TO NEW YORK IN WHICH YOU GAVE IT AS 41.47 50.20 IF SUCH WERE

over

Marconigram 5 (Page two):

CORRECT SHE WOULD BE IN HEAVY FIELD ICE AND NUMEROUS BERGS HOPE AND TRUST MATTERS ARE NOT AS BAD AS THEY APPEAR REGARDS.

HAINS.

◀ *Olympic–Mesaba* Marconigram
The *Mesaba* reports a lot of ice and icebergs on the morning of 15 April. (Mike Beatty collection.)

▶ **Account of Wages**
Accounts of wages like this *Olympic* example were issued to all surviving *Titanic* crewmen, itemising the wages they earned during the *Titanic*'s incomplete maiden voyage. 'We lost everything except what we had on and a change at home,' fireman Charles Judd wrote not long after being rescued by the *Carpathia*. 'All my poor mates are gone to the bottom – nearly all Shirley [Southampton] chaps on our watch. I do not know what pay we shall get, as we have only five days due to us, and we cannot claim any more. As soon as the ship strikes our pay stops – hard lines; but never mind, I am only too thankful to be where I am.' (George Behe collection.)

▲ Brown Brothers Glass Negative

On the evening of 18 April 1912, the *Carpathia* arrived in New York with 712 survivors on board. That same night the *Titanic*'s lifeboats were dropped off at Pier 59 – all that was left of the great ship. The next day, on this glass-plate negative, a photographer for Brown Brothers captured an iconic image of the lifeboats being inspected at the pier. (Mike Beatty collection.)

AFTERMATH

▲ Postcard
A card showing the graves of *Titanic* victims whose recovered bodies were brought to the city of Halifax by the funeral ships. Many of the victims' bodies remain unidentified to this very day. (George Behe collection.)

▲ *Titanic* Deck Chair Slat
The cable steamer *Minia* not only recovered bodies of the *Titanic* victims but also pieces of floating wreckage. This *Titanic* backrest slat was originally part of a damaged deck chair that was picked up by the recovery vessel. (John Lamoreau collection.)

▶ Masthead Light
The masthead light of the cable ship *Mackay-Bennett*, which was involved in the recovery of the *Titanic*'s dead. The ship was eventually scrapped in 1965. (Kalman Tanito collection.)

▼ Wreck Wood
After the *Titanic*'s sinking, ships were sent out to the scene to recover the bodies of those who died in the disaster. At the same time, the sailors recovered pieces of floating wreck wood and later made different items from it, ranging from simple cribbage boards to ornate boxes and chess tables. This is one such example. (Kalman Tanito collection.)

▶ **_Titanic_ Story Solicitation Postcard**
Newspapers were desperate for any news they could get about the sinking. This rare postcard shows one method of how they tried to obtain a scoop for their publication. (Mike Beatty collection.)

◀ **The Dublin Review**
It's not often discussed just how much cargo was lost on board the _Titanic_. One such loss appears to have been crates of the April 1912 edition of _The Dublin Review_, which subsequently had to be reprinted and sent overseas again according to this cover. (Mike Beatty collection.)

▶ **Salesman Sample Postcard**
This salesman card solicited sellers for memorial items produced by the W.M. Prilay Post Card Company. Many companies capitalised on public interest in the disaster. (Mike Beatty collection.)

"STEAMER TITANIC" Largest and most luxurious in the World. Launched at Belfast, Ireland, May 1911. Length 882 ft. 6 inches. Displacement 66,000 tons. On her maiden trip struck a mammoth iceberg on Sunday, April 14th at 10.25 P. M. 41o 46 minutes, north latitude 50o 14 minutes, west longitude. The worst disaster known in Marine History. Sunk at 2.20 A. M. April 15, 1912, with a loss of over 1300 lives.

THIS SPACE MAY BE USED FOR WRITING

THIS SPACE FOR ADDRESS ONLY

We have large picture of Titanic size 15x22, picture also shows map and location of disaster and brief description of same. Extra heavy stock sells for .25 cost you .10 how many do you want?

We also sell cards like this sample assorted colors, two different views .50 per 100.

**W. M. Prilay Post Card Co.
Pittsfield, Maine**

Pub. by W. M. Prilay, Pittsfield, Me.

▲▶ **Advertising Postcards**
Postcards like these examples served to drum up interest in hastily assembled post-disaster books that consisted mainly of transcribed newspaper articles about the *Titanic*'s sinking. (George Behe collection.)

▲ Postcard
One of the first things the White Star Line did after the disaster was recall all its *Olympic/Titanic* advertising postcards and issue revised versions with the *Titanic*'s name carefully deleted from the new cards. On this card, note the empty space to the right of the *Olympic*'s name, which in former days contained the name of her younger sister. (George Behe collection.)

▲ Memorial Postcards

Meanwhile, independent publishers were coming out with hurriedly produced *Titanic* memorial postcards of widely varying quality. Our first two examples look as if they were created by amateur artists, while the French card (bottom right) is of marginally better quality. (George Behe collection.)

▲ **Memorial Postcard**
A German post-disaster postcard mistakenly depicts the *Titanic* sinking by the stern instead of by the bow. (George Behe collection.)

◀ **Memorial Postcard**
This post-disaster card depicts a fanciful version of the *Titanic*'s collision with the iceberg. In reality, no gaping hole like the one shown here ever existed; instead, damage to the ship was restricted to the starboard side of her hull, below the waterline, and permitted water to enter the ship's forward compartments through small gaps that opened between separated hull plates. (George Behe collection.)

FRAGMENTS OF HISTORY: THE SHIP 147

▲ **Deceptive Postcards**
Meanwhile, other opportunistic postcard manufacturers were eager to cash in on the *Titanic*'s name even though they had nothing genuine to offer their customers. Instead, the publishers substituted illustrations of other ships on their postcards and falsely labelled them as showing the *Titanic*. The first three examples have the Cunard Line's *Lusitania/Mauretania* being forced to fill in for the lost White Star liner, while our fourth card shows the *Olympic* fulfilling that same function. (George Behe collection.)

◀ **Deceptive Postcards**

Not being satisfied with creating fake *Titanic* memorial postcards, unscrupulous publishers even issued fake *Carpathia* postcards. Our first example shows SS *Momus* falsely representing the *Carpathia*, but that's nothing compared to our second postcard showing the huge *Lusitania/Mauretania* with three of her four funnels painted out to 'resemble' the much smaller, single-funnelled *Carpathia*. (George Behe collection.)

▶ **Photograph**
The rescue ship *Carpathia* became well known to the public because of the part she played in rescuing the *Titanic*'s survivors. Several professional photographers subsequently created crude painted backdrops depicting the rescue ship so that people could pay to have their photographs taken in front of it. (George Behe collection.)

▲ **Rivet Punchings**
Various divots, also known as rivet punchings, that were turned into commemorative items after the *Titanic* disaster. (Kalman Tanito collection.)

▲ **Memorial Plaque**
Many different types of memorial items were created in the wake of the *Titanic* disaster. The one pictured here is a circular glass frame with marine snail shells arranged lengthwise around its perimeter. A memorial *Titanic* postcard is presented underneath the frame's glass along with more seashells and some dried plant material. (George Behe collection.)

◀ **Postcards**

During the two government inquiries into the *Titanic* disaster, it was determined that the Leyland liner *Californian* was within sight of the *Titanic* during the entire time the big White Star liner was sinking. Despite subsequent conspiracy theories alleging that a so-called 'mystery ship' lay directly between the *Titanic* and *Californian* that night, no credible evidence of such a vessel's existence has ever been produced. (George Behe collection.)

▲ Postcard

This postcard picturing a generic view of the *Olympic* and *Titanic* was issued by the American Bankers Association to promote the use of its travellers' cheques. After the disaster, the association was undoubtedly glad it hadn't thought to include the *Titanic*'s name anywhere on the card. (George Behe collection.)

▲ **Bremaeker Medallion**
This 2in by 2½in bronze relief medallion was designed and engraved by Eugene De Bremaeker for the Society of Friends in Holland and Belgium as a Medal of Art shortly after the *Titanic* sinking. The medallion pays tribute to the invention of wireless and the men who operated it that made rescue possible for survivors of both the *Titanic* and other recent sinkings. (Mike Beatty collection.)

FRAGMENTS OF HISTORY: THE SHIP   155

◀▼ **Balham and Tooting Medallion**
Volunteers who worked for the Balham and Tooting relief fund were presented with this aluminium medallion. (Mike Beatty collection.)

◀ **Relief Fund Pin**
A small celluloid pin most likely given out to contributors to one of the *Titanic* relief funds. (Mike Beatty collection.)

⏪ **Magic Lantern Slides**

Top Image: A.J. Clapham Magic Lantern Slides
Clapham was one of many magic lantern slide producers to capitalise on the disaster. A theatre owner could buy a set of forty black and white slides for $10 or colour slides for $20. (Mike Beatty collection.)

Bottom Image: Levi Company Magic Lantern Slides
Another lantern slide manufacturer to offer a *Titanic* disaster set for theatres was the Levi Company. (Mike Beatty collection.)

▶ **Myriorama Show**
This elaborate sinking show in England was presented by Charles W. Poole. Myriorama shows featured a large moving painted canvas on rollers. Emotional narratives and special effects were added to the changing scenes on the canvas to create a spectacular show. (Mike Beatty collection.)

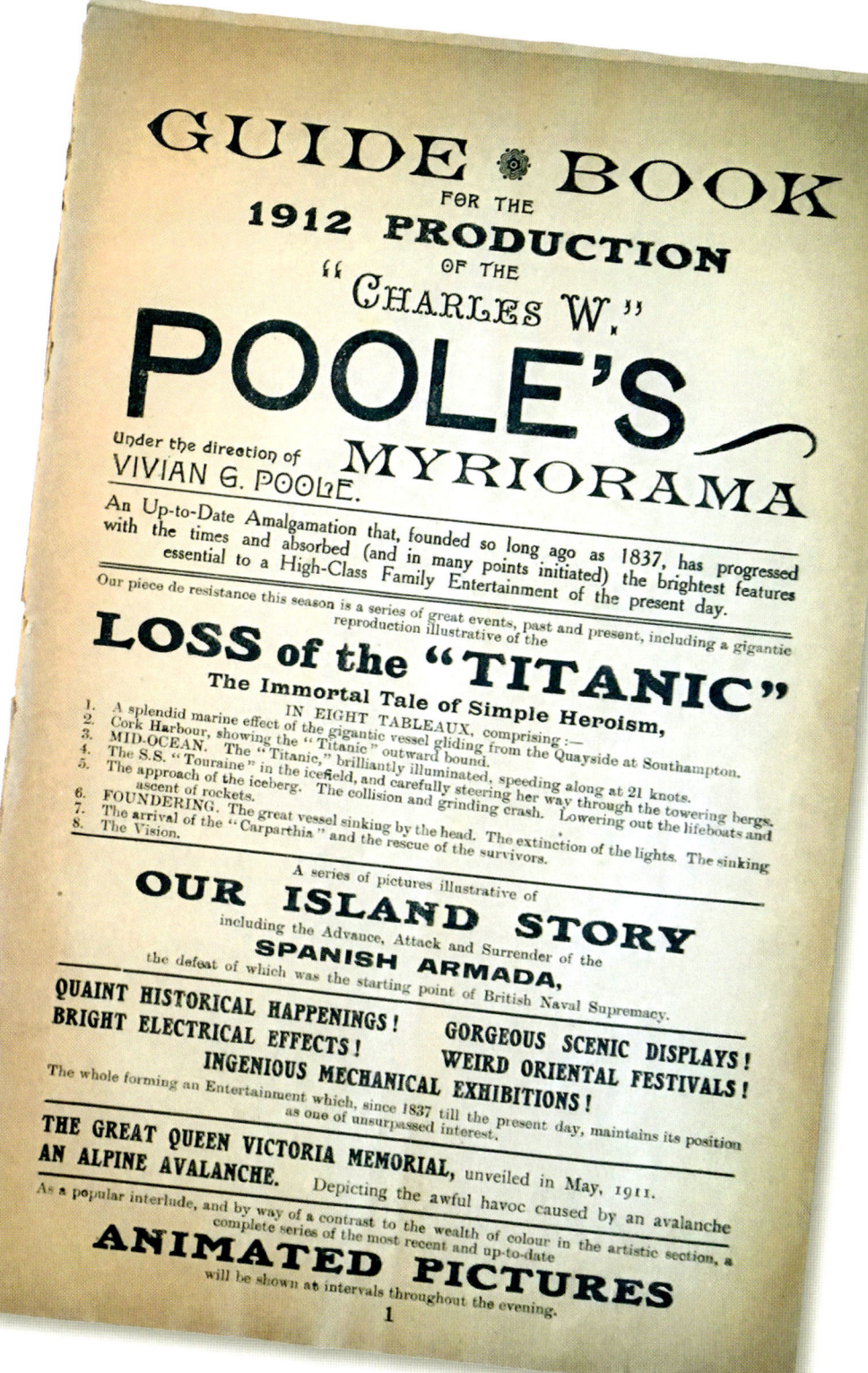

▲▲ **Slide Show on Riverboat**
This leaflet advertised another lantern slide show of the sinking, this time located in an unusual venue. The show was on board the riverboat *Greater New York* floating theatre. (Mike Beatty collection.)

▲ **Lantern Show Leaflet**
A leaflet advertising a magic lantern slide show of the sinking at the G. Roberts & Son's Picture Palace in England shortly after the disaster. (Mike Beatty collection.)

▲ **New York City Attractions**
This weekly booklet publication of attractions and events going on in New York City was already advertising a *Titanic* sinking magic lantern show on the week of 29 April. (Mike Beatty collection.)

▲ **Victor Records, 'Nearer My God to Thee'**
This was a popular hymn even before the *Titanic* disaster. This record pressing from about 1905 on the Victor label would have been played on early Victrolas. (Mike Beatty collection.)

FRAGMENTS OF HISTORY: THE SHIP   161

◀ **Edison Disc, 'Nearer My God to Thee'**
Edison's flat disc players could only play his proprietary version of flat records and not standard 78s. Here's a copy of the famous hymn from around 1915 made to play on his machines. (Mike Beatty collection.)

▶ **Edison Disc, 'Sinking of the *Titanic*'**
Here's another Edison flat disc featuring a popular post-sinking tribute. (Mike Beatty collection.)

▲ 'Be British' 78 RPM Record
Another popular tribute was this UK release.
(Mike Beatty collection.)

FRAGMENTS OF HISTORY: THE SHIP   163

▲ **Edison Roll 'Sinking of the *Titanic*'**
If you were still using an older Edison phonograph that played rolls, you could get a copy of this song for those players. (Mike Beatty collection.)

◀▲ **Piano Roll, 'Wreck of the *Titanic*'**
If you had a player piano at home, tribute music was available in that format as well. (Mike Beatty collection.)

▶ **Fred Pansing Oilette**
Marine painter Fred Pansing was commissioned by the White Star Line to paint an advertising piece for the *Titanic* and *Olympic* to be displayed in its offices and ticket agencies. Pansing completed this work not long before he died in 1912. After the sinking, the oilette was modified to show more lifeboats and was redistributed throughout the company. This oilette shows the version with lifeboats added. (Mike Beatty collection.)

▲ **Apple *Titanic* Postcard**
This prize-winning display made from whole and dried slices of Graventein apples was featured at the third-annual Graventein Apple Show held in Sebastopol, California, from 19 to 25 August 1912. It was created by William S. 'Scottie' Liddle. (Mike Beatty collection.)

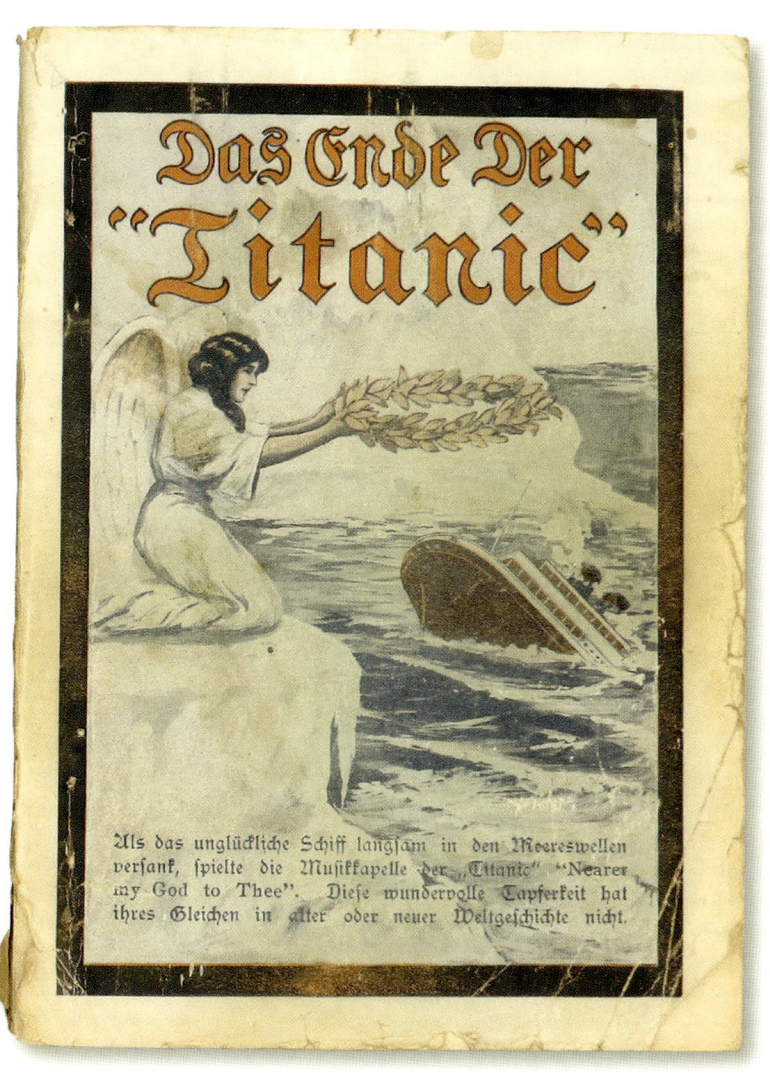

▲ **German Books**
German-language *Titanic* books from soon after the disaster, published in Chicago, Illinois. (Mike Beatty collection.)

◀ **Polish Book**
An extremely rare Polish book printed soon after the *Titanic* disaster. It was published in North Chicago, Illinois. (Mike Beatty collection.)

FRAGMENTS OF HISTORY: THE SHIP 169

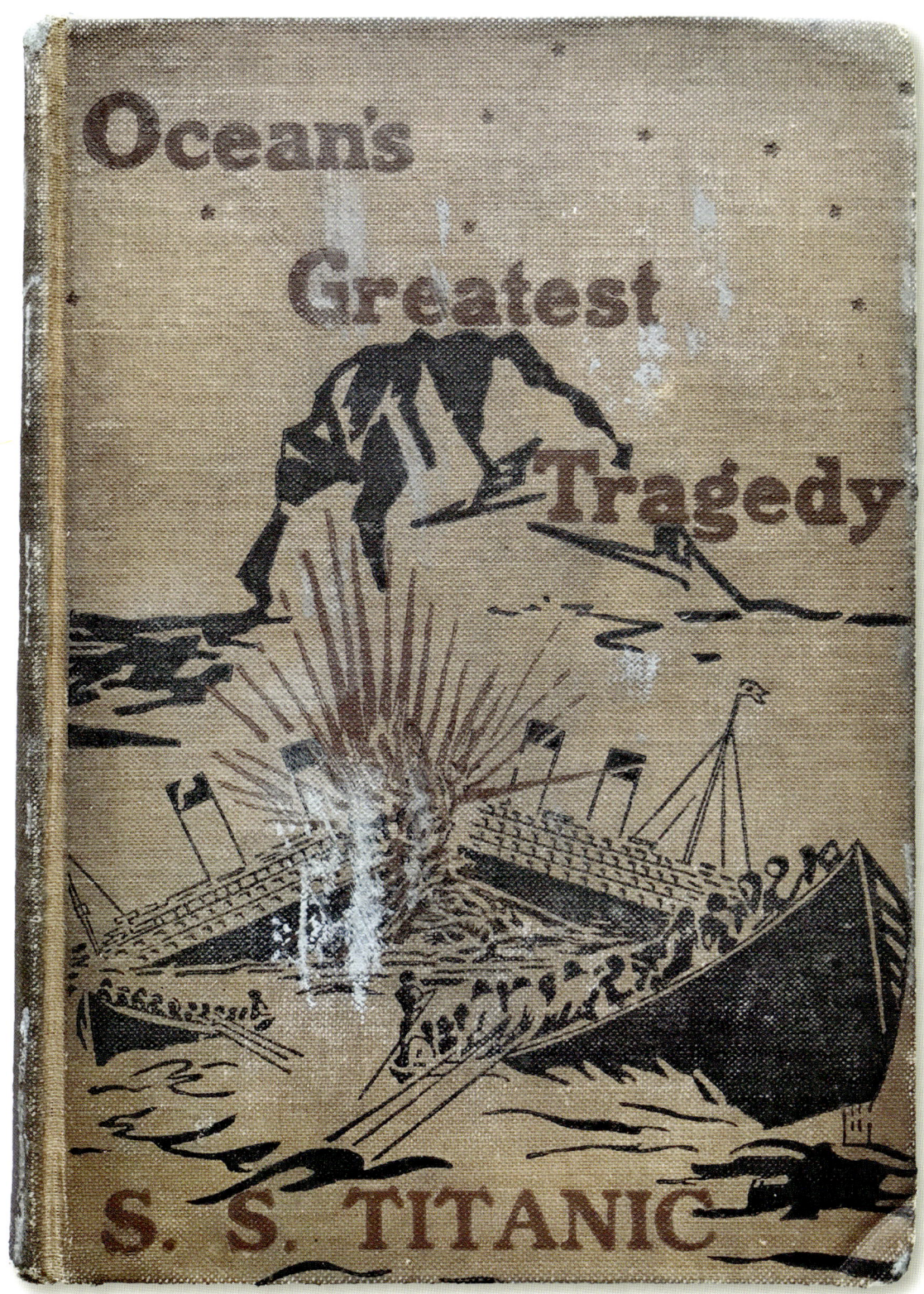

▶ *Titanic* Book
Another extremely rare book about the *Titanic* disaster, published in English in Chicago, Illinois. (George Behe collection.)

◀ **Medal**
Captain Prothero's Liverpool Shipwreck and Humane Society medal, which he received for his heroism when the *Carpathia* was torpedoed on 15 July 1918. (Kalman Tanito collection.)

▶ **Movie Poster**
The Clifton Webb/Barbara Stanwyck film *Titanic* debuted in 1953 and helped create a permanent *Titanic* interest in at least one co-author of this present book. (George Behe collection.)

AUTHOR BIOGRAPHIES

MIKE BEATTY was born, raised and still lives in the Philadelphia region with his wife Jessica and daughter Mira. Having had an early interest in history, he started reading about the *Titanic* around 1983 and his interest only grew from there. In 1987, he was lucky enough to meet six survivors; around the same time, he started collecting items related to the *Titanic*, *Olympic* and *Britannic*, a hobby that he still actively pursues. In 2017, he wrote the book *Sincerely Harry: The Letters of Henry Tingle Wilde*, a compilation of letters written by the *Titanic*'s Chief Officer Wilde during his career; twenty-eight of these letters are in Mike's collection. Mike currently runs the Facebook page 'White Star Memorabilia', where collectors display their original items and provide helpful advice to steer buyers away from fake items that are so prevalent today. He is currently a Titanic International trustee and contributor to their journal *Voyage*. His other interests include restoration: he has restored antique cars and radios and has been restoring a 1906 house for fifteen years, which has earned him the local Historic District Preservation Award in 2010. He works in the power industry as a controls and instrumentation technician and is also a part-time antiques dealer.

GEORGE BEHE became interested in the *Titanic* as a small boy in the 1950s when he found a 1912 book about the disaster on his grandmother's bookshelf. One of the book's illustrations depicted a dying swimmer raising his arm in futile supplication towards passengers in a lifeboat that was being rowed right past him, which made George wonder how he might have felt if he and his mother had been in that lifeboat and the dying swimmer were his father. After completing his hitch in the army in 1972, George began researching the disaster in earnest and eventually attended a Titanic Historical Society convention, where he met several *Titanic* survivors in person. He also met Don Lynch, with whom he began exchanging research information, and with whom he

eventually made research trips to England and The National Archives. George became friends with several of the *Titanic* survivors including Winnifred Quick Van Tongerloo, who happened to live only a few miles from his home. He is a prolific author and contributor and has written numerous articles for the Titanic Historical Society as well as several books for The History Press, including *Fate Deals a Hand*, *Voices from the Carpathia* and *On Board RMS Titanic*. George is now retired and spends as much time as possible researching the *Titanic* disaster and sharing his newest discoveries.

JOHN LAMOREAU lives with his wife in rural Oregon, surrounded by children and grandchildren. His interest in history and ships comes naturally: his genealogy includes Myles Standish (9th great-grandfather), who sailed on the *Mayflower*, and Andre L'Amoureux (8th great-grandfather), a ship's captain who escaped persecution in France and arrived in the American colonies in 1701. From an early age, Lamoreau was fascinated by the stories of the *Titanic*. As a collector of historical documents, his interest in the *Titanic* was further inspired after obtaining a handwritten letter from passenger Francis D. Millet, which was the start of his collection. An educator, Lamoreau brought his interest in the *Titanic* to his classroom: one class project became a cover story for the Titanic Historical Society's journal *The Titanic Commutator* ('Edith Rosenbaum: World War One Letter from France', Issue #218). Lamoreau has also teamed up with a local chef to host a popular annual recreation of the last first-class dinner served on the *Titanic*. Called 'a taste sensation', the dinner offers guests every item served that night and has attracted out-of-country guests including relatives of the *Titanic's* passengers. He enjoys sharing his collection with others, as well as collaborating with other researchers and family members of passengers.

DON LYNCH was born in Coeur d'Alene, Idaho, and raised in Washington State, where he began researching the *Titanic* while still in high school in Spokane. In the five decades since, he has travelled to museums and archives throughout the United States, Canada, England and Ireland to conduct his research. He has met and interviewed twenty passengers from the *Titanic*, as well as numerous relatives of survivors and victims. For many years he has been the official historian for the Titanic Historical Society. In 1992, Don wrote the text for the book *Titanic: An Illustrated History*, which went on to spend twelve weeks on the *New York Times*' bestseller list. Director James Cameron hired Don as the historian for the 1997 movie *Titanic* and also as a consultant on his 2003 *Ghosts of the Abyss* project, where they dove to the *Titanic* in a Russian submersible to film the wreck for Cameron's 3D, large-format documentary. Don wrote the text for the companion book and was a contributor for Cameron's book *Exploring the Deep: The Titanic Expeditions*.

TREVOR POWELL lives in Philadelphia, Pennsylvania, where he attended college at Temple University. His interest in maritime history began at the age of 7 after reading a book on the *Titanic* disaster in elementary school. He began collecting ocean liner memorabilia twenty years ago and has a primary focus on the Olympic-class liners as well as the *Lusitania*. His other interests include genealogy and historic architectural preservation. Trevor is a member of the Titanic International Society.

KALMAN TANITO lives in Budapest, Hungary, with his wife and two sons. As a child, he moved with his parents to Finland, where he became interested in the *Titanic* and researched the many Scandinavian passengers on the ship. This resulted in correspondence with several experts in the field, as well as meetings with survivors at different conventions. He has also published several articles in the Titanic Historical Society's magazine, *The Titanic Commutator*. As fate would have it, when James Cameron's *Titanic* came out in 1997, Tanito happened to sit next to a woman whose relative didn't survive the disaster. This led to a research trip to northern Finland with his friend Juha Peltonen, where they met family members of other passengers. With Phillip Gowan, a fellow researcher, good friend and best man at his wedding, he followed the *Titanic* trail in the Balkans. He has specialised in rare publications about the **Titanic** catastrophe, especially in non-English ones, but he is also interested in other ocean liners of the past. In his spare time, he is a keen photographer and spends time on modern cruise ships, where he is able to mix these two passions.

BY THE SAME AUTHORS

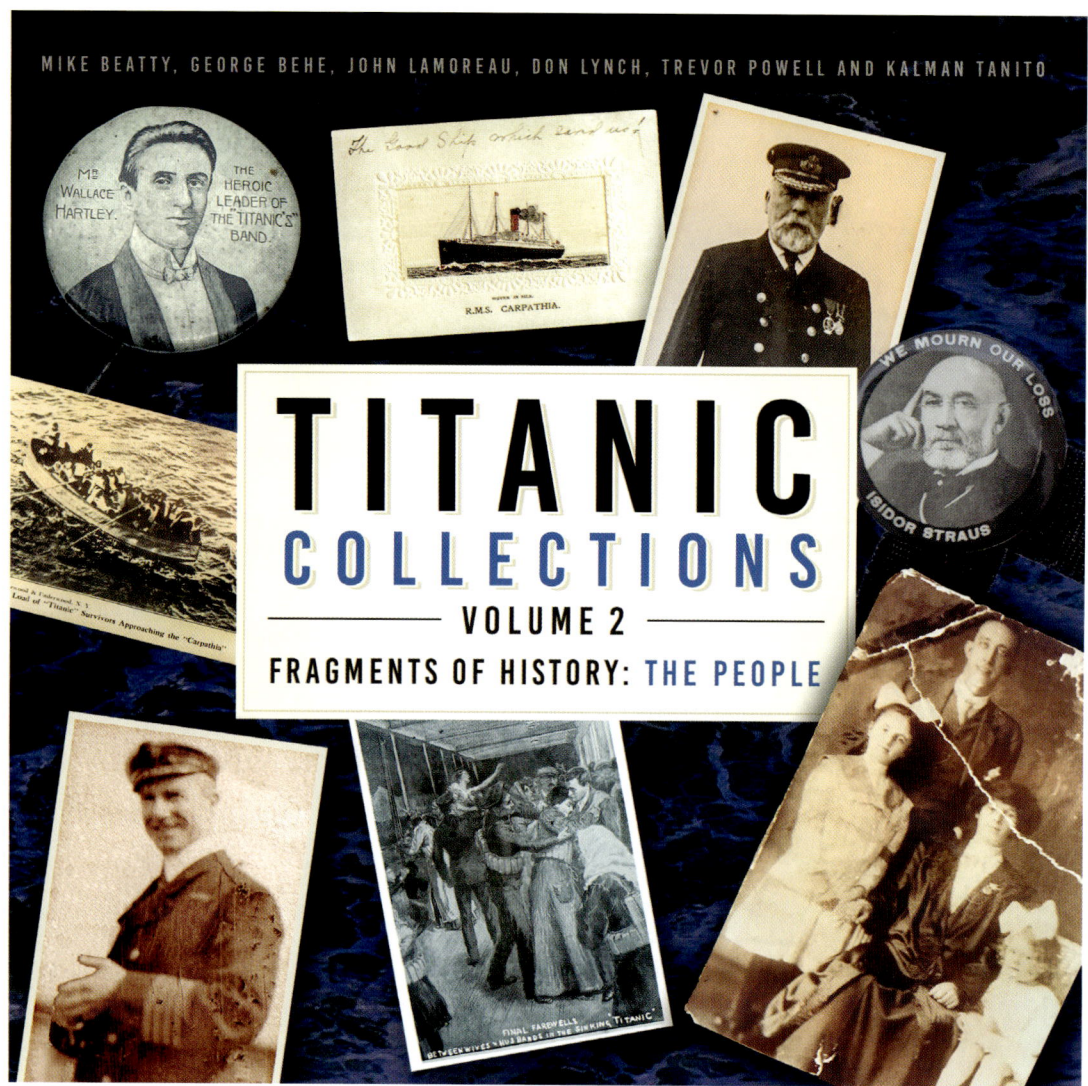

978-1-80399-334-8

The destination for history
www.thehistorypress.co.uk